Mathematical Wisdom in Everyday Life

From Common Core to Math Competitions

Kevin Wang
Kelly Ren
John Lensmire

PUBLISHED BY ARETEEM INSTITUTE
WWW.ARETEEM.ORG
ALL RIGHTS RESERVED.

ISBN: 1-944863-02-8
ISBN-13: 978-1-944863-02-9

First printing, September 2016.

Contents

Introduction 5

1   Number Sense 9

1.1   Key Concepts 10

1.2   Exercises - One: Basic Number Sense Problems 13

1.3   Exercises - Two: Number Sense Problems with More Depth 20

1.4   Exercises - Three: Challenging Number Sense Problems 31

2   Ratio, Proportion, and Percentage 41

2.1   Key Concepts 42

2.2   Exercises - One: Basic Percentage Problems 50

2.3   Exercises - Two: Percentage Problems with More Depth 55

2.4   Exercises - Three: Challenging Percentage Problems 62

3   Chickens and Rabbits 71

3.1   Key Concepts 72

3.2 **Exercises - One: Basic Chicken & Rabbit Problems** 73

3.3 **Exercises - Two: Chicken & Rabbit Problems with More Depth** . 80

3.4 **Exercises - Three: Challenging Chicken & Rabbit Problems** 87

4 Motion—Speed, Time, and Distance 95

4.1 **Key Concepts** 97

4.1.1 Motion in Flowing Water 98

4.1.2 Relative Speed and Frame of Reference 100

4.1.3 Meet Up, Farewell, and Catch Up 101

4.1.4 Problem Solving Strategies for Motion Problems 102

4.2 **Exercises - One: Basic Motion Problems** 105

4.3 **Exercises - Two: Motion Problems with More Depth** 110

4.4 **Exercises - Three: Challenging Motion Problems** 114

5 Work Related Problems 121

5.1 **Key Concepts** 122

5.2 **Exercises - One: Basic Work Problems** 125

5.3 **Exercises - Two: Work Problems with More Depth** 129

5.4 **Exercises - Three: Challenging Work Problems** 134

Answer Keys to Problems 139

Index 157

Introduction

This book is a part of the ongoing effort by Areteem Institute to inspire students, parents, and teachers to gain a deeper understanding and appreciation of mathematics. The content is organized to emphasize on the proper implementation of the Common Core Mathematics Standard, focusing on conceptual understanding, problem solving, and real world applications. Some contest (AMC8, MATHCOUNTS, and ZIML Division M) level problems are included to challlenge talented middle school students or advanced elementary school students.

The book is divided into five chapters. The first two chapters cover fractions, decimals, and conversions as well as ratios, proportions, and percentages. These chapters review middle school Common Core Standards, particularly those in the domains of RP, NS, and EE at the 6th, 7th, and 8th grade levels. They also cover more advanced topics such as converting repeating decimals to fractions. The next three chapters cover real life applications of the material from the first two chapters, divided by category into "chicken and rabbit questions", "motion questions", and "work-related questions".

As we are talking about the mathematical wisdom in everyday life, we introduce an everyday family, a typical American family, the Winchers, and hopefully you can relate and incorporate the mathematical wisdom into your own family life. The Winchers are a family made up of four people; John and Mary, the parents, Connor, a junior in high school, and Samantha, a 7th grader. Their lives are typical, just like any other family, and their everyday life is peppered with mathematics. John Wincher is a businessman at a local business, and Mary Wincher is a full time mom who volunteers at her kids' schools and tutors students online. Connor is an honors student who is aiming to become a scientist or a doctor. Samantha enjoys all her subjects in school, and is also more inclined towards math and science. This family appreciates mathematics as a basic element in the world. Here, we will see various types of math problems worked into their regular schedules. Would you be able to enjoy mathematics the same way as they do in your everyday life?

Each chapter of this book starts with a summary of the math concepts in correspondence to the Common Core Standards, accompanied by a number of examples, and followed by 60 exercise problems, grouped into three sections of increasing challenge levels. These problems are designed to help students enhance their mastery of the math concepts and practice to solve real world problems in their daily lives.

The book is available as a Student Workbook as well as a Solutions Manual. The Student Workbook contains the summary of the math concepts, examples, exercises, and answer keys to all the exercise problems. The Solutions Manual includes detailed solutions to all of the exercise problems. Most questions include multiple solutions. Students are encouraged to find multiple ways of solving each question. For many of the examples and exercise problems, both non-algebraic and algebraic solutions are included to demonstrate the pros and cons of using algebra. Algebra is not required to solve the problems in this book. Clever combinations of logic and arithmetic are sufficient to solve the problems. For students who do have experience in algebra, the comparison would show that it is often beneficial *not* to set up equations at first; instead, they should use logical reasoning and basic arithmetic operations to try and solve the problems without using any variables. The ideas used in non-algebra solutions usually lead to very clever fast solutions, which are also less error-prone. More importantly, it is a good thought-provoking exercise to develop creative logical thinking skills. Teachers and parents are encouraged to give students plenty of time and opportunities to figure out their own ways of solving the problems before presenting any of the solutions to them. Only after they spend a lot of effort to think, the students will be able to appreciate more of the solutions provided and absorb the methods in more depth.

Teachers and students that are in 6th, 7th, and 8th grade math can use this book to teach/learn mathematical reasoning and problem solving, fraction operations, and real-world applications. The start of each chapter summarizes the specific Common Core standards emphasized in the chapter.

About Areteem Institute

Areteem Institute is an educational institution that develops and provides in-depth and advanced math and science programs for K-14 (Elementary School, Middle School, High School, and two years of college) students and teachers. Areteem programs are accredited supplementary programs by the Western Association of Schools and Colleges (WASC). Students may attend the Areteem Institute through these options:

- Live and real-time face-to-face online classes;
- Self-paced classes by watching the recordings of the live classes;
- Summer Intensive Camps and Winter Boot Camps.

The Areteem courses are designed and developed by educational experts and industry professionals to bring real world applications into the STEM education. The programs are ideal for students who wish to win in Math Competitions (AMC, AIME, USAMO, IMO, ARML, MathCounts, Math Olympiad, ZIML, etc.), Science Fairs (County Science Fairs, State Science Fairs, national programs like Intel Science and Engineering Fair, etc.) and Science Olympiad, or purely want to enrich their academic lives by taking more challenges and developing outstanding analytical, logical thinking and creative problem solving skills.

Since 2004 Areteem Institute has been teaching with the methodology that is highly promoted by the new Common Core State Standards: stressing the conceptual level understanding of the math concepts, problem solving techniques, and solving problems in the real world applications. With the guidance from experienced and passionate professors, students are motivated to explore concepts deeper by identifying an interesting problem, researching it, analyzing it, and using a critical thinking approach to come up with multiple solutions.

Hundreds of math students who have been trained at Areteem achieved top honors and earned top awards in major national and international math competitions, including Gold Medalists in the International Math Olympiad (IMO), top winners at the USA Math Olympiad (USAMO/JMO), top winners at the Zoom International Math League (ZIML), and top winners at the MathCounts National. Many Areteem Alumni have graduated from high school and gone on to enter their dream colleges such as MIT, Cal Tech, Harvard, Stanford, Yale, Princeton, U Penn, Harvey Mudd College, UC Berkeley, UCLA, etc. Those who have graduated from college are now playing important roles in their fields of endeavor.

Further information about Areteem Institute, as well as updates and errata of this book, can be found online at http://www.areteem.org.

Acknowledgments

This book contains many years of collaborative work by the staff of Areteem Institute. This book could not have existed without their efforts. Huge thanks go to the Areteem staff for their contributions.

The examples and problems in this book were either created by the Areteem staff or adapted from various sources, including other books and online resources. It is not practical to list all such resources here. We extend our gratitude to the original authors of all these resources.

Special thanks are owed to Adelle Wang, who created the story of the Winchers family, and baked those mouth-watering pretzels! They were so delicious! (See Example 5.1 on Page 123)

1. Number Sense

The concepts introduced in this chapter directly correspond to Common Core Math Standards as shown in the following table.

6th Grade	6.NS.1, 6.NS.2, 6.NS.3, 6.NS.4
7th Grade	7.NS.1, 7.NS.2, 7.NS.3, 7.EE.3, 7.EE.4
8th Grade	8.EE.7

In addition to the standards above, problems and concepts in this section will help strengthen understanding of the following domains.

6th Grade	6.NS, 6.EE
7th Grade	7.NS, 7.EE
8th Grade	8.EE

1.1   Key Concepts

Multiple

Counting by a number (starting with 0) gives all the multiples of that number. For example,
$$0, 2, 4, 6, 8, 10, \ldots$$
are the multiples of 2 while
$$0, 5, 10, 15, 20, 25, \ldots$$
are the multiples of 5.

Example 1.1

List the 5 smallest multiples of 7.

Solution

The 5 smallest multiples of 7 are $0, 0 + 7 = 7, 7 + 7 = 14, 14 + 7 = 21$ and $21 + 7 = 28$, so they are $0, 7, 14, 21, 28$.

Division with Remainder

When dividing two numbers, they do not always divide equally. The partial answer is called the quotient and any leftover is called the remainder.

Example 1.2

Find the quotient and remainder when 22 is divided by 5. ♣

Solution

Note that 20 is a multiple of 5 with 2 left over. Therefore the remainder is 2. Since $4 \times 5 = 20$, we have the quotient is 4.

Example 1.3

Write $22 \div 5$ as (i) an improper fraction, (ii) a mixed number, and (iii) a decimal. ♣

Solution

(i): We simply have $22 \div 5 = \dfrac{22}{5}$.

(ii): Remember from above we know that $22 \div 5$ is 4 with remainder 2. Hence, $22 \div 5 = 4 + \dfrac{2}{5} = 4\dfrac{2}{5}$.

(iii): Using long division

$$
\begin{array}{r}
4.4 \\
5\,\overline{)\,22.0} \\
20 \\
\hline
20 \\
20 \\
\hline
0
\end{array}
$$

so we have that $22 \div 5 = 4.4$.

Factor

If a number goes evenly into another number, we call it a factor. For example, $8 \div 4 = 2$ so 4 is a factor of 8. We can also say that 8 is divisible by 4.

Remark

Another common name for factor is divisor. Thus, the phrase "4 is a divisor of 8" has the same meaning as "4 is a factor of 8".

Example 1.4

Find all the factors of 10.

Solution

We can write $10 = 1 \times 10 = 10 \times 1$ and also $10 = 2 \times 5 = 5 \times 2$. Since there is no other way to write 10 as the product of two whole numbers, $1, 2, 5, 10$ are all the factors of 10.

Example 1.5 Factor Pairs

Multiplication is commutative, so the order we multiply two numbers does not matter. Therefore, we can pair up factors of a number. In the previous example, 10 had factors $1, 2, 5, 10$. As $1 \times 10 = 2 \times 5 = 10$ we pair up 1 with 10 and 2 with 5.

Remark

Be careful! Sometimes factors will be paired with themselves. For example, 4 has factors $1, 2, 4$. In this case $1 \times 4 = 2 \times 2 = 4$, so 2 must be paired with itself.

Prime Number

A whole number is a prime number if its only factors are one and itself.

Rational Number

A number is a rational number if it is result of dividing two integers. For example $3 \div 5 = \dfrac{3}{5}$ is a rational number.

Example 1.6

The following numbers

$$2, -5, \frac{8}{-3}, 0.25$$

are all rational numbers. This is because

$$2 = \frac{2}{1}, -5 = \frac{-5}{1}, \frac{8}{-3} = \frac{-8}{3}, 0.25 = \frac{1}{4}$$

so each number can be written in the required form.

1.2   Exercises - One: Basic Number Sense Problems

Example 1.7

Simplify the fraction $\dfrac{10}{40}$.

Solution

Since 10 is a factor of 40, we know we can divide the numerator and denominator by 10 to simplify the fraction. We have

$$\frac{10}{40} = \frac{10 \div 10}{40 \div 10} = \frac{1}{4}.$$

Example 1.8

Add the fractions $\frac{1}{5} + \frac{3}{5}$.

Solution

Since both fractions have the same denominator, we just need to add the numerators of the fractions. We therefore have

$$\frac{1}{5} + \frac{3}{5} = \frac{1+3}{5} = \frac{4}{5}$$

as our final answer.

Proper Fraction, Improper Fraction, and Mixed Number

- A proper fraction is a fraction whose numerator is less than the denominator. The value of a proper fraction is less than 1. For example, the numbers $\frac{1}{2}$ and $\frac{23}{79}$ are proper fractions.
- An improper fraction is a fraction whose numerator is greater than the denominator. The value of an improper fraction is greater than 1. For example, the numbers $\frac{3}{2}$ and $\frac{64}{17}$ are improper fractions.
- A mixed number is a combination of a whole number and a proper fraction. For example, the numbers $2\frac{1}{2}$ and $12\frac{3}{7}$ are mixed numbers.

Example 1.9

Multiply $\dfrac{8}{5} \times \dfrac{7}{5}$. Write your answer as a mixed number.

♣

Solution

To multiply fractions we just need to multiply the numerators and denominators of the fractions separately. We have

$$\frac{8}{5} \times \frac{7}{5} = \frac{8 \times 7}{5 \times 5} = \frac{56}{25}.$$

Note that $25 \times 2 = 50$, so we can write $56 \div 25 = 2$ with remainder 6. Hence,

$$\frac{56}{25} = 2 + \frac{6}{25} = 2\frac{6}{25}$$

is our answer written as a mixed number.

Example 1.10

Convert the fraction $\dfrac{3}{8}$ to a decimal.

♣

Solution

To convert a fraction to a decimal, we divide the numerator by the denominator:

$$3 \div 8 = 0.375.$$

Therefore the decimal form of $\dfrac{3}{8}$ is 0.375.

Example 1.11

Convert the decimal number 2.95 to (1) a mixed number; (2) an improper fraction. ♣

Solution

(1) We first convert the part 0.95 to a fraction. To do this, we note that there are two decimal places after the decimal point, so the denominator is 100 (in general, we count the number of decimal places, and then put the same number of "0"s following the "1"), thus

$$0.95 = \frac{95}{100};$$

next we simply the fraction to the lowest terms:

$$\frac{95}{100} = \frac{5 \times 19}{5 \times 20} = \frac{19}{20};$$

finally, we add the integer part 2 to make it a mixed number: $2\frac{19}{20}$.

(2) To convert 2.95 to an improper fraction, we follow the steps of (1) to convert it to a mixed number first: $2.95 = 2\frac{19}{20}$, then combine the whole number part "2" into the fraction:

$$2.95 = 2\frac{19}{20} = 2 + \frac{19}{20} = 2 \times \frac{20}{20} + \frac{19}{20} = \frac{40}{20} + \frac{19}{20} = \frac{59}{20}.$$

Therefore the improper fraction form of 2.95 is $\frac{59}{20}$.

Remark

We will discuss operations with fractions in more depth later in this chapter after we have introduced greatest common factors and least common multiples in more detail.

Problem 1.1 Find all the factors of 15.

Problem 1.2 How many factors does the number 30 have?

Problem 1.3 How many factors does the number 16 have?

Problem 1.4 List all the factors of 120 less than or equal to 10.

Problem 1.5 Simplify the fraction $\dfrac{3}{36}$.

Problem 1.6 Find the quotient and remainder when $35 \div 3$.

Problem 1.7 Write the improper fraction $\dfrac{41}{4}$ as a mixed number.

Problem 1.8 Write the improper fraction $\dfrac{39}{6}$ as a simplified mixed number.

Problem 1.9 Write the mixed number $4\dfrac{5}{7}$ as an improper fraction.

Problem 1.10 What is $\dfrac{8}{19} + \dfrac{11}{19}$?

Problem 1.11 Calculate $\dfrac{5}{24} + \dfrac{7}{24}$. Write your answer as a simplified fraction.

Problem 1.12 Calculate $6 - \dfrac{7}{3}$. Write your answer as an improper fraction.

Problem 1.13 What is $\dfrac{15}{11} + \dfrac{19}{11}$? Write your answer as a mixed number.

Problem 1.14 What is the sum of the mixed numbers $2\frac{1}{8}$ and $4\frac{3}{8}$? Write your answer as a mixed number with a simplified fraction.

Problem 1.15 Multiply the fractions $\frac{3}{5} \times \frac{1}{4}$.

Problem 1.16 Multiply and simplify $10 \times \frac{4}{5}$.

Problem 1.17 What is $\frac{5}{7} \times \frac{3}{2}$? Write your answer as a mixed number.

Problem 1.18 Convert the decimal 0.02 to a simplified fraction.

Problem 1.19 Convert $\frac{7}{20}$ into a decimal.

Problem 1.20 Write the decimal 11.5 as an improper fraction.

1.3 Exercises - Two: Number Sense Problems with More Depth

Common Factor

Given two numbers, any whole number that is a factor of both numbers is called a common factor.

Example 1.12

The factors of 15 are
$$1, 3, 5, 15$$
and the factors of 12 are
$$1, 2, 3, 4, 6, 12.$$
Therefore the common factors of 15 and 12 are 1 and 3.

Greatest Common Factor

The greatest common factor of two numbers is the largest common factor of the two numbers. Greatest common factor is often abbreviated as GCF.

Example 1.13

In the previous example, we saw that $1, 3$ were the common factors of 15 and 12. Therefore the GCF of 1 and 3 is 3.

Remark

Remember from the previous sections that "divisor" has the same meaning as "factor". Similarly, the term "greatest common divisor" or "GCD" is the same as the "greatest common factor" or "GCF".

Common Multiple

Given two numbers, any positive whole number that is a multiple of both numbers is called a common multiple.

Example 1.14

The multiples of 3 are

$$3, 6, 9, 12, 15, 18, 21, 24, 27, 30, 33, 36, 39, \ldots$$

and the multiples of 4 are

$$4, 8, 12, 16, 20, 24, 28, 32, 36, 40, 44, \ldots.$$

Therefore, the common multiples of 3 and 4 are $12, 24, 36, \ldots.$

Least Common Multiple

The least common multiple of two numbers is the smallest common multiple of the two numbers. Least common multiple is often abbreviated as LCM.

Example 1.15

In the previous example, we saw that $12, 24, \ldots$ were the common multiples of 3 and 4. Therefore the LCM of 3 and 4 is 12.

Example 1.16

Describe all the common multiples of 2 and 3.

Solution

The multiples of 2 are

$$2, 4, 6, 8, 10, 12, 14, 16, 18, 20, 22, 24, 26, \ldots$$

and the multiples of 3 are

$$3, 6, 9, 12, 15, 18, 21, 24, 27, \ldots.$$

Therefore the least common multiple is 6 and the common multiples are

$$6, 12, 18, 24, \ldots.$$

Note that the common multiples are all multiples of 6 (the least common multiple). Therefore we can say that the common multiples of 2 and 3 are all multiples of 6.

Remark

> The result of the previous example holds in general: the common multiples
> of two numbers are exactly the multiples of the least common multiple.

We earlier saw how to reduce fractions when the numerator was a factor of the denominator or vice versa. In general, all we need is common factors to help reduce fractions.

Example 1.17

Reduce the fraction $\dfrac{15}{25}$.

Solution

15 is not a factor of 25. However, if we look at the factors of 15, which are

$$1, 3, 5, 15$$

and the factors of 25, which are

$$1, 5, 25$$

we see that 5 is the greatest common factor of 15 and 25. Dividing the numerator and denominator by 5 we get

$$\frac{15}{25} = \frac{15 \div 5}{25 \div 5} = \frac{3}{5},$$

a simplifed version.

We can use common factors and greatest common factors to simplify or reduce fractions, but how do we know when they are fully simplified?

Fraction in Lowest Terms

We say a fraction is written in lowest terms if the greatest common factor of its numerator and denominator is 1.

Example 1.18

The fraction $\dfrac{3}{5}$ is in lowest terms because 3 has factors

$$1, 3$$

and 5 has factors

$$1, 5$$

so the greatest common factor of $3, 5$ is 1.

Example 1.19

The fraction $\dfrac{9}{15}$ is not in lowest terms because it can be reduced to

$$\frac{9}{15} = \frac{9 \div 3}{15 \div 3} = \frac{3}{5}.$$

We have just seen that $\dfrac{3}{5}$ is in lowest terms.

Remark

Generally the phrases "fraction in lowest terms", "simplified fraction", or "reduced fraction" all have the same meaning.

Common multiples are very useful for adding and subtracting fractions. We have already reviewed how to add and subtract fractions when they have the same denominator.

If fractions do not have the same denominator, they first need to be rewritten so they share a *common denominator*. Often the product of the two denominators is used as the common denominator, as in the following example:

Example 1.20

To add

$$\frac{3}{10} + \frac{7}{20}$$

we multiply the numerator and denominator of the first fraction by 20 (the second denominator) and similarly multiply the numerator and denominator of the second fraction by 10 (the first denominator). This gives us

$$\frac{3 \times 20}{10 \times 20} + \frac{7 \times 10}{20 \times 10} = \frac{60}{200} + \frac{70}{200} = \frac{130}{200}.$$

Note however, that the fraction $\frac{130}{200}$ can easily be simplified to $\frac{13}{20}$. Further we see that this method can easily lead to fairly large numbers.

Note that the common denominator

$$10 \times 20 = 200$$

is a common multiple of 10 and 20. However, as 10 is a factor of 20, the LCM of 10 and 20 is just 20, and we can use this as our common denominator:

Example 1.21

To add

$$\frac{3}{10} + \frac{7}{20}$$

we first note that the LCM of 10 and 20 is 20 so we can use 20 as our common denominator. Since $20 = 10 \times 2$ and $20 = 10 \times 1$ we have

$$\frac{3}{10} + \frac{7}{20} = \frac{3 \times 2}{10 \times 2} + \frac{7 \times 1}{20 \times 1} = \frac{6}{20} + \frac{7}{20} = \frac{13}{20}$$

as our answer.

Note here that the numbers we needed to work with are much smaller, and that in this case our fraction was already in lowest terms.

Remark

You may still need to simplify the answer when adding fractions, even if you use the LCM as the common denominator. When the lowest common multiple of the denominators is used as the common denominator, it is called the *lowest common denominator*. It is generally a good practice to use lowest common denominators for calculations, because the numbers are smaller and easier to work with.

Example 1.22

Calculate $\dfrac{5}{6} - \dfrac{1}{4}$.

Solution

We first calculate the LCM of 6 and 4. Multiples of 6 are

$$6, 12, 18, 22, \ldots$$

and multiples of 4 are

$$4, 8, 12, 16, 20, \ldots$$

so the least common multiple is 12. We then have

$$\frac{5}{6} - \frac{1}{4} = \frac{5 \times 2}{6 \times 2} - \frac{1 \times 3}{4 \times 3} = \frac{10}{12} - \frac{3}{12} = \frac{7}{12}.$$

Note that 7 and 12 share only 1 as a factor, so this fraction is in lowest terms.

Example 1.23

Connor and Samantha read separate books next to each other. Connor can read one chapter in 14 minutes while it takes Samantha 21 minutes to read a chapter. If they have less than an hour to read together, how long should they read so that when they stop they both have read a whole number of chapters in their books?

Solution

Connor reads one chapter in 14 minutes. Therefore he will finish a chapter after every multiple of 14 minutes. Similarly, Samantha will finish a chapter after every multiple of 21 minutes. We look at multiples of 14 less than 60,

$$14, 28, 42, 56$$

and multiples of 21 less than 60

$$21, 42,$$

as they only have 1 hours or 60 minutes to read together. The least common multiple is 42, and this is the only common multiple of 14 and 21 less than 60. Thus, Connor and Samantha should read together for 42 minutes.

Remark

In the previous example, Connor and Samantha both read for 42 minutes. However, this means they read a different number of chapters. Connor reads

$$42 \div 14 = 3 \text{ chapters,}$$

while Samantha reads

$$42 \div 21 = 2 \text{ chapters.}$$

Problem 1.21 List the common factors of 16 and 28.

Problem 1.22 What is the greatest common factor of 48 and 60?

Problem 1.23 List the first 5 common multiples of 2 and 4.

Problem 1.24 Find the least common multiple of 10 and 15.

Problem 1.25 Write the fraction $\dfrac{20}{25}$ in lowest terms.

Problem 1.26 Write the fraction $\dfrac{24}{36}$ in lowest terms.

Problem 1.27 Write the decimal 0.48 as a fraction in lowest terms.

Problem 1.28 Convert the fraction $\dfrac{5}{8}$ to a decimal.

Problem 1.29 Calculate the sum: $\dfrac{5}{12} + \dfrac{1}{6}$.

Problem 1.30 Calculate $\dfrac{1}{4} - \dfrac{1}{10}$.

Problem 1.31 What is the sum of $\dfrac{1}{10}$ and $\dfrac{1}{15}$? Write your answer as a simplified fraction.

Problem 1.32 What is $\dfrac{1}{3} \times \dfrac{9}{4}$? Write your answer as a simplified fraction.

Problem 1.33 Calculate $3\dfrac{1}{2} \times 2\dfrac{2}{3}$ and write your answer as a simplified improper fraction.

Problem 1.34 Calculate $\dfrac{5}{12} \div \dfrac{10}{3}$. Express your answer as a fraction in lowest terms.

Problem 1.35 Divide $3\frac{3}{8}$ by 3. Write your answer as a mixed number.

Problem 1.36 Mr. and Mrs. Smith are both teachers. Mr. Smith's class has 24 students, while Mrs. Smith's class has 30 students. Mr. Smith divides his class into N groups, with each group having the same size. Mrs. Smith also divides her class into N groups, with each group having the same size. What is the largest value of N so this is true?

Problem 1.37 Billy has 60 green marbles and 40 blue marbles. He wants to put the marbles in bags so that each bag as the same number of green and blue marbles. Billy wants to use as many bags as possible, but wants to make sure that each bag has the same number of green and blue marbles. How many bags should Billy use?

Problem 1.38 Nick and Justin run laps around the school. Nick completes a lap every 4 minutes while Justin completes a lap every 6 minutes. They agree to run until they meet up for a second time at the starting line. How many minutes in total do they run?

Problem 1.39 Grace and Andrew enjoy building towers with blocks. Grace has blocks that are 10 cm high and Andrew has blocks that are 6 cm high. What is the smallest height that both Grace and Andrew can build with they blocks?

Problem 1.40 Jane's family loves to come visit her. Her parent's come visit every 6 weeks, and her grandparents come visit every 8 weeks. How often do her parents and grandparents both visit the same week?

1.4 Exercises - Three: Challenging Number Sense Problems

The ideas and methods in previous sections can be combined to create more challenging questions, which are explored in this section. The problem solving techniques are often the same, but the problems may require additional steps to solve.

Example 1.24

Samantha takes her younger cousins Billy and Tommy to buy a piece of candy at the store. The only money Billy brought is a few quarters and the only money Tommy brought is a few dimes. Billy tries to buy a piece of candy, but the clerk says, "Sorry Billy, you need 18 more cents to buy the candy." To show off, Tommy tries to buy the same piece of candy, and the clerk says, "You don't have enough money either Tommy, you need 8 more cents. Samantha bought the candy herself with a $1 bill and gave it to her cousins to share. How much did the piece of candy cost?

Solution

Samantha paid with a $1 bill, so we know the candy is less than $1 or 100 cents. Since Billy has only quarters, he has either

$$25, 50, \text{ or } 75$$

cents. If he is 18 cents short of buying the candy, the price of the candy (in cents) must be one of

$$25 + 18 = 43, 50 + 18 = 68, 75 + 18 = 93$$

---

cents. Tommy has only dimes so the amount of money he has is

$$10, 20, 30, 40, 50, 60, 70, 80, \text{ or } 90$$

cents. He is 8 cents short of buying the candy, so the candy must be

$$18, 28, 38, 48, 58, 68, 78, 88, \text{ or } 98$$

cents. Hence, the candy must be 68 cents, as this is the only price that fits with both possibilities.

Remark

It is good to practice converting word problems into math questions to help understand really what the word problem is asking. The following example is the word problem above reworded as a pure math question.

Example 1.25

The whole number N is between 1 and 100. We know that N leaves remainder 18 when divided by 25 and leaves remainder 8 when divided by 10. What is N?

Solution

Note that the work in the solution to the previous example applies to this question as well. Therefore, $N = 68$.

The previous sections have included problems converting fractions to decimals and decimals to fractions. These were done in the case of decimals with one, two, or three decimal places, so much of the work was simplifying fractions, as, for example, the writing 0.55 as a fraction amounts to simplifying the fraction $\dfrac{55}{100}$.

We end this section by presenting a method for converting repeating decimals into fractions.

Example 1.26

Convert the fraction $\frac{1}{3}$ to a decimal.   ♣

Solution

Using long division we have

$$
\begin{array}{r}
0.33 \\
3\,\overline{\smash{)}\,1.00} \\
\underline{0.9} \\
10 \\
\underline{9} \\
1
\end{array}
$$

From here, note that the pattern of dropping the 0 to get 10, noting that $9 = 3 \times 3$ is the smallest multiple of 3 less than 10, and subtracting to get

$$10 - 9 = 1,$$

will repeat over and over. Hence $\frac{1}{3}$ as a decimal will be $0.3333\cdots$ where the 3's repeat forever. We use the notation $0.\overline{3}$ to denote that the 3 keeps repeating.

We can use an algebraic trick to help convert repeating fractions to decimals. Let's look at the previous example in reverse.

Example 1.27

Convert the repeating decimal $0.333333\cdots = 0.\overline{3}$ into a fraction.   ♣

Solution

Of course we already know the answer is $\frac{1}{3}$, but we present a general method.

We are used to letting x be an unknown in math problem, although the number is not really unknown, here let $x = 0.\overline{3}$. We will be able to write an equation and "solve" for x to show $x = \dfrac{1}{3}$. We have that

$$x = 0.333333\cdots = 0.\overline{3},$$

so multiplying by 10 on both sides we get

$$10x = 3.33333\cdots = 3.\overline{3}.$$

Notice that both $0.\overline{3}$ and $3.\overline{3}$ have the same decimal parts. Thus, if we subtract the two equations we get

$$10x - x = 3.33333\cdots - 0.333333\cdots = 3.\overline{3} - 0.\overline{3}. = 3$$

Therefore,
$$9x = 3$$

so
$$x = \frac{3}{9}.$$

Simplifying (as 3 is a factor of 9),

$$\frac{3}{9} = \frac{3 \div 3}{9 \div 3} = \frac{1}{3},$$

so $0.\overline{3} = \dfrac{1}{3}$ as expected.

Example 1.28

Convert $0.2\overline{5}$ to a fraction.

Solution

We begin as in the previous example and let $x = 0.2\overline{5}$. We also multiply by 10 on both sides, so that since

$$x = 0.2555555\cdots = 0.2\overline{5}$$

---

we have
$$10x = 2.55555\cdots = 2.\overline{5}.$$

Unlike the previous example, the decimal parts here do not match. Multiplying by 10 again we have
$$100x = 25.5555\cdots = 25.\overline{5},$$

so now the decimal portion of $100x$ and $10x$ match. Subtracting them we get
$$100x - 10x = 25.5555\cdots - 2.55555\cdots = 25.\overline{5} - 2.\overline{5} = 23.$$

Therefore
$$90x = 23$$

so
$$x = \frac{23}{90}.$$

Hence we can write the decimal $0.2\overline{5} = \dfrac{23}{90}$ as a fraction.

As one last example, we show how the method presented above can be used to explain the often misunderstood argument that the repeating decimal $0.999999\cdots = 0.\overline{9}$ is equal to 1.

Example 1.29

Show that $1 = 0.999999\cdots = 0.\overline{9}$. ♣

Solution

We are used to letting x be an unknown in a problem, but here let $x = 0.\overline{9}$. We will be able to write an equation and "solve" for x to show $x = 1$. Since we know that
$$x = 0.999999\cdots = 0.\overline{9},$$

multiplying by 10 on both sides we get
$$10x = 9.99999\cdots = 9.\overline{9}.$$

Notice that both $0.\overline{9}$ and $9.\overline{9}$ have the same decimal parts. Thus, if we subtract the two equations we get

$$10x - x = 9.99999\cdots - 0.999999\cdots = 9.\overline{9} - 0.\overline{9} = 9.$$

Therefore,

$$9x = 9$$

so

$$x = \frac{9}{9} = 1.$$

Since we also know that $x = 0.\overline{9}$ we must have that

$$0.\overline{9} = 1$$

as claimed!

Problem 1.41 Find the largest whole number less than 40 that leaves a remainder of 2 when divided by 7.

Problem 1.42 Find the greatest 2-digit number that leaves a remainder of 4 when divided by 11.

Problem 1.43 Find the smallest whole number that leaves a remainder of 2 when divided by 5 and a remainder of 4 when divided by 6.

Problem 1.44 How many of the improper fractions

$$\frac{20}{1}, \frac{20}{2}, \frac{20}{3}, \frac{20}{4}, \ldots, \frac{20}{20}$$

can be simplified to whole numbers?

Problem 1.45 Josh goes shopping for socks. Each pair of socks costs $8. Josh buys as many socks as he can with $100. How many pairs of socks does Josh buy? What is his change?

Problem 1.46 Emily Elizabeth is painting rainbow eggs for Easter. She painted Red, Orange, Yellow, Green, Gray, Blue, and Purple colors in order for a palette of 80 eggs delivered by Farmer Brown. What is the color of the last egg?

Problem 1.47 Four friends, Alice, Bob, Charlie, and Drew, count to 50, as in the chart below:

A	B	C	D
1	2	3	4
8	7	6	5
9	10	11	12
...	14	13

Which of the four friends says the number 50?

Problem 1.48 George wrote a computer program. If you type in a positive integer, George's program outputs the sum of all the positive factors of that integer. For example, if you type in 6, the program outputs 12, because the factors of 6 are $1, 2, 3, 6$ and $1 + 2 + 3 + 6 = 12$. Suppose George's friend Paul types in 7 and then runs the program. If Paul types in this result and runs the program again, what is the output?

Problem 1.49 Consider the number 15. Multiply together all the factors of 15. What is the result?

Problem 1.50 Frank's goal is the drink 5 liters of water per day. He drinks water out of a bottle that holds $\frac{2}{3}$ of a liter of water. How many bottles of water does Frank need to drink to reach his goal of 5 liters? Express your answer as a mixed number.

Problem 1.51 Julie buys lemons and limes at the grocery store. The price of lemons is 4 lemons per \$1 and the price of limes is 6 limes for \$1. Julie buys a total of 6 lemons and 21 limes. How much money does Julie spend in total?

Problem 1.52 Tom is planting a tree and needs to buy soil. He knows he needs $3\frac{2}{3}$ cubic feet of soil and the soil costs \$3.50 dollars per cubic foot. How much money will Tom need to spend on the soil? Express your answer as a mixed number.

Problem 1.53 Florence has some candy bars. If she keeps 3 for herself, she can distribute the remaining candy bars evenly among 4 children. If she keeps 5 for herself, she can distribute the remaining candy bars evenly among 9 children. If Florence has less than 30 candy bars, exactly how many does she have?

Problem 1.54 Convert the fraction $\dfrac{2}{9}$ to a decimal.

Problem 1.55 Express the fraction $\dfrac{5}{11}$ as a repeating decimal.

Problem 1.56 Write the fraction $\dfrac{5}{6}$ as a decimal.

Problem 1.57 Convert the decimal $0.\overline{6}$ into a fraction.

Problem 1.58 Write the repeating decimal $0.2\overline{6}$ as a simplified fraction.

Problem 1.59 Write the repeating decimal $0.\overline{21}$ as a fraction.

Problem 1.60 Write the number $2.\overline{2}$ as a mixed number.

2. Ratio, Proportion, and Percentage

The Winchers family loves to go hiking. Their favorite hiking trail is one that winds through a beautiful mountain and passes by a waterfall, then comes down to a sparsely traveled road by the ocean. One time, in the afternoon, as they were walking along this road, Samantha observed a man climbing a ladder near a telephone pole to presumably fix the wiring at the top. As she watched, the man swarmed up the ladder, reaching the top of the pole and reaching for the wires. She wondered: how far did he have to climb to reach the top?

As it was the afternoon, the sun happened to be in a place where everyone's shadows were long. Could Samantha have used the proportion of her height to her shadow's length to find out how far the telephone pole man climbed?

The concepts introduced in this chapter involve ratio, proportion, and percentage. These concepts directly correspond to Common Core Math Standards as shown in the following table.

6th Grade	6.RP.1, 6.RP.2, 6.RP.3, 6.EE.7
7th Grade	7.RP.2, 7.RP.3, 7.NS.3, 7.EE.2, 7.EE.3, 7.EE.4
8th Grade	8.EE.7

In addition to the standards above, problems and concepts in this section will help strengthen understanding of the following domains.

6th Grade	6.RP, 6.NS, 6.EE
7th Grade	7.RP, 7.NS, 7.EE
8th Grade	8.EE, 8.F

2.1 Key Concepts

Ratio

Ratio is a relation between two quantities of the same kind, expressed as "a to b" or $a : b$, where a represents the first quantity and b the second quantity. Sometimes a ratio is expressed as a quotient of the two numbers, indicating how many times the first number contains the second, with value $\frac{a}{b}$, not necessarily a whole number.

Example 2.1

Connor's science class has 20 students and 1 teacher. We would say that the student to teacher ratio in his class is $20 : 1$. We could also say that the teacher to student ratio is $1 : 20$.

Proportion

Proportion is the comparison of two pairs of ratios. Two ratios are proportional if they have equal values when both are reduced to the simplest form.

Example 2.2

Suppose that all 7 of Connor's classes have 20 students and 1 teacher per class. Then in total there are

$$7 \times 20 = 140$$

students and

$$7 \times 1 = 7$$

teachers. Hence the student to teacher ratio is $140 : 7$. The ratio $140 : 7$ can be reduced to $20 : 1$, so we can say the school maintains a $20 : 1$ student to teacher ratio. In this case, the ratios $140 : 7$ and $20 : 1$ are proportional.

In mathematical notation, if two ratios $a : b$ and $c : d$ are proportional, we write

$$a : b = c : d \qquad \text{or} \qquad \frac{a}{b} = \frac{c}{d}.$$

Percentage

A percentage is a number or ratio expressed as a fraction of 100, denoted using the symbol "%".

Much of what we already learned about ratios and proportions apply to percentages because the percentage is a ratio of a number to 100. Percentages are often used as a convenient way of comparing the values of different ratios.

Percentage is very common in real life scenarios. For example, if we say 20 percent or 20% of the population in the area are kids, it means that out of every 100 people, 20 are kids. Other common examples of percentage are stores running a promotion by giving a percentage discount or banks offering percentage interest rates for savings accounts.

Example 2.3

Mary goes shopping at the Sell-It store. The store is running a promotion of "50% Off" of everything at the store. This means that if you buy an item that is originally priced at $100, now you save $50, only paying $50. How much does Mary save if she buys an item that is originally priced at $48?

Solution

The total discount is 50%, so Mary saves

$$50\% \times 48 = 0.50 \times 48 = 24$$

dollars on the item.

Basic number sense is very useful for dealing with ratios and percentages. We often want to reduce ratios to "1 to something". For example, when we talk about a teacher to student ratio, we usually refer to how many students compare to one teacher.

Example 2.4

In Hudson High School, there are 120 teachers and 4800 students, so the teacher-student ratio is 120 : 4800. How many students are there for one teacher?

Solution

We want to reduce the ratio from 120 : 4800 to 1 to something. Many students will start by using long division to divide

$$4800 \div 120 = 40,$$

so we see that

$$120 : 4800 = (120 \div 120) : (4800 \div 120) = 1 : 40,$$

so there are 40 students for one teacher. In most situations, the long division process is tedious and error-prone. It is often easier and faster to cancel out common factors from both sides. In this case, if you recognize that 120 is a factor of 4800, great! Otherwise, you can always use smaller factors to help reducing the ratio. For example, it is much easier to see that 10 is a factor of both sides, so

$$120 : 4800 = (120 \div 10) : (4800 \div 10) = 12 : 480.$$

From here it is easier to see that 12 is a factor of 480, so we have

$$12 : 480 = (12 \div 12) : (480 \div 12) = 1 : 40,$$

so we still arrive at the same answer. With a little practice, you will learn to recognize common factors quickly, and with this improved number sense you will be able to reduce ratios without using long division.

It is important to learn how to quickly reduce ratios so we can tell when ratios are proportional.

Example 2.5

Are the following ratios proportional?
  (1) $3 : 4$ and $24 : 32$
  (2) $2 : 3$ and $4 : 9$
  (3) $7 : 9$ and $28 : 36$

Solution

  (1) Since the numbers 24 and 32 have a common factor of 8, we may reduce the ratio

$$24 : 32 = (24 \div 8) : (32 \div 8) = 3 : 4,$$

  and therefore the two ratios are proportional.

  (2) Both ratios are already reduced to the lowest terms, so since they are not equal they are not proportional. To double check, note that 9 is a multiple of 3, so we have

$$2 : 3 = (2 \times 3) : (3 \times 3) = 6 : 9,$$

  and we see that the ratios $2 : 3 = 6 : 9$ and $4 : 9$ are not equal.

---

(3) Since the numbers 28 and 36 have a common factor of 4, we may reduce the ratio

$$28 : 36 = (28 \div 4) : (36 \div 4) = 7 : 9,$$

and therefore the two ratios are proportional.

Example 2.6

Find x in the following proportions.
(1) $12 : 17 = x : 68$
(2) $1.68 : 12 = 7 : x$
(3) $95 : 6.5 = x : 5.2$

Solution

(1) A typical procedural solution is to perform "cross-multiplication": the product of the inner terms equals the product of the outer terms, as

$$12 : 17 = x : 68 \text{ so } \frac{12}{17} = \frac{x}{68}$$

so

$$17 \times x = 12 \times 68$$

and x can be found by carry out the multiplication and division.

However, a student with better number sense can recognize that 17 is $\frac{1}{4}$ of 68.

Therefore, using proportions, 12 should also be $\frac{1}{4}$ of x. This immediately gives the answer

$$x = 4 \times 12 = 48,$$

without the need for large multiplications and divisions.

(2) Numbers with decimal points are more complicated to handle and cross-multiplication involving decimals can get very messy. One of the possible starting steps is to get rid of the decimal points by enlarging the numbers in one or both of the ratios proportionally so that the numbers involved are whole numbers:

$$1.68 : 12 = (1.68 \times 100) : (12 \times 100) = 168 : 1200.$$

Then the problem becomes

$$168 : 1200 = 7 : x.$$

Here, the ratio $168 : 1200$ can be reduced to simpler terms after both 168 and 1200 are divided by 8:

$$168 : 1200 = (168 \div 8) : (1200 \div 8) = 21 : 150.$$

The proportion $21 : 150 = 7 : x$ is already much easier to solve. However, we may still reduce the ratio $21 : 150$ further if we notice that 21 and 150 are both divisible by 3:

$$21 : 150 = (21 \div 3) : (150 \div 3) = 7 : 50.$$

Aha! Look at the proportion now: $7 : 50 = 7 : x$. The value of x is already found with $x = 50$.

An alternative solution is to recognize that

$$1.68 = 0.24 \times 7,$$

so then by the definition of proportion,

$$12 = x \times 0.24,$$

and we can solve to get

$$x = 12 \div 0.24 = 50$$

as before.

(3) This problem also involves some pesky decimal values. Enlarging the second numbers in both ratios 10 times, we get a new proportion $95 : 65 = x : 52$. It is easy to notice that 95 and 65 both have factor 5, so we can reduce $95 : 65 = 19 : 13$. At this point the proportion in question becomes $19 : 13 = x : 52$, and it takes some good number sense to see that 13 is a factor of 52. Therefore we have

$$19 : \frac{13}{13} = x : \frac{52}{13},$$

which becomes

$$19 : 1 = x : 4.$$

Now the numbers are small enough and it is a good time to apply the cross-multiplication method: $x = 19 \times 4 = 76$.

Example 2.7

(1) Samantha wanted to buy a new iPhone case that was originally priced $28, and she had a coupon for 15% off. What would be the discounted price of the iPhone case, after the coupon was applied?

(2) In the same situation above, if the state tax rate was 8%, what was the final amount Samantha had to pay?

(3) The cashier, Roger, applied the tax first and got the amount Samantha needed to pay before Samantha showed him the coupon. Roger took the coupon and applied the 15% discount and reached a final amount for Samantha to pay. Samantha made the payment and received her new lemon-patterned iPhone case. She was very happy and put it on the phone immediately. However, on her way home, she kept wondering if she might have ended up paying more tax because the cashier applied the tax before applying the coupon. Can you help her figure it out?

Solution

(1) To calculate the discounted price, we calculate the discount amount by multiplying the original price by the percent discount, which is $28 \times 15\%$, and then subtract this value from the original price. This method can be simplified to the following expression and value: $28 \times (1 - 15\%) = 28 \times 0.85 = 23.80$ dollars.

(2) To pay the price with sales tax added, we should calculate the tax first, and then add it to the price to pay. The discounted price was already calculated to be $23.80, all we need to do now is to add the 8% tax onto the value: $23.08 \times (1 + 8\%) = 23.80 \times 1.08 = 25.704$, so Samantha would pay $25.70 after rounding.

(3) The question is asking to compare the results of two procedures: (I) Add the sales tax first, then apply the discount coupon on the new (higher) amount; (II) Apply the coupon first, then add the sales tax on the new (lower) amount.

In the solution for the previous question, we have calculated the resulting price 25.70 of procedure (I). In order to compare the two procedures, we would need to calculate the resulting price of procedure (II)... or would we? The calculations involving these numbers are quite complicated.

> ### Remark
>
> Rule of thumb regarding calculations: Delay complicated calculations as much as possible!

For the purpose of comparison, let us write down the mathematical expressions for the two procedures. For procedure (I), we add the tax first, and then apply the discount:

$$28 \times (1 + 8\%) \times (1 - 15\%).$$

For procedure (II), we apply the discount first, and then add the tax:

$$28 \times (1 - 15\%) \times (1 + 8\%).$$

By the commutative property of multiplication, these two expressions have the exact same value. Therefore the resulting prices are the same for both procedures. In this problem, we see that critical thinking is much more important than following given procedures. By writing out the mathematical expression, we solved the problem without performing any calculation!

Problem Solving Strategies

For problems involving percentage in general, two methods may be employed.
- Assume that one of the amounts involved is 100 (or another convenient nonzero value), and identify the other amounts relative to the assumed amount.
- Make a table for the different amounts involved and list out the known amounts and unknown amounts, and find the relationship of the amounts and represent them accordingly.

Example 2.8

Conner was taking an investment course at a local community college. In the course he was given a practice account with imaginary currency, and a list of imaginary companies for investment. During the first class period, he invested equal amounts from the account on stocks of three of the imaginary companies AFZ, BLS, and CIW. The AFZ stock never changed value for the entire course. The BLS stock increased by 10% by the end of the second class, and then decreased by 10% of the new value by the end of the third class. The CIW stock decreased by 20% by the end of the second class, and then increased by 20% of the new value by the end of the third class. At the end of the third class, which stock had the highest value, and which had the lowest?

Solution

It is tempting to say there was no change in value for each of the stocks by the end of the third class. However, we shall perform some calculations and see if that is true.

Assume Conner invested $100 on each of the stocks at the beginning. The following table shows the value change of each stock at each class:

Stock	1st class	2nd class	3rd class
AFZ	100	100	100
BLS	100	$100 \times (1 + 10\%) = 110$	$110 \times (1 - 10\%) = 99$
CIW	100	$100 \times (1 - 20\%) = 80$	$80 \times (1 + 20\%) = 96$

From the table, it is clear that the AFZ stock has the highest value and CIW has the lowest.

2.2 Exercises - One: Basic Percentage Problems

In this section, we present a few examples involving basic percentage. Exercise problems follow afterwards.

Example 2.9

A coffee mug marked $10 was on sale for $6. What is the rate of discount in percentage? ♣

Solution

The amount of discount is

$$10 - 6 = 4$$

---

dollars. Thus the rate of discount is

$$4 \div 10 = 0.4 = 40\%.$$

Example 2.10

A salesman who works on a commission basis earns 12% of his sales. How much was his commission on a $265 sale?

Solution

The commission is the sale amount times the commission rate. Hence the answer is

$$265 \times 12\% = 265 \times 0.12 = 31.80$$

dollars.

Example 2.11

Samantha has exactly 2 pennies, 2 nickels, one dime and one quarter in her pocket. What percent of a dollar is in her pocket?

Solution

Since one dollar is 100 cents, it is easier if all the values of the coins are converted to cents as well. A penny is worth 1 cent, a nickel 5 cents, a dime 10 cents, and a quarter 25 cents, hence Samantha has

$$2 \times 1 + 2 \times 5 + 1 \times 10 + 1 \times 25 = 47$$

cents in her pocket. Therefore Gina has

$$47 \div 100 = 0.47 = 47\%$$

of a dollar.

Problem 2.1 Last month the price of gas was $1.10 per gallon. This month gas is selling for $1.32 per gallon. Find the percentage increase in the price of gas per gallon.

Problem 2.2 There are 180 days in a school year. John was present 85% of the total days. How many days was he present?

Problem 2.3 Sam received a grade of 80% on a geometry test. If he solved 24 problems correctly, how many problems were on the test?

Problem 2.4 Jim receives a weekly salary of $200. He spends $60 per week on gas. What percent of his weekly salary does he use for gas?

Problem 2.5 Last year Areteem Institute had an enrollment of 500 students. This year the enrollment is 800 students. What is the percent of increase in student enrollment?

Problem 2.6 Jess paid $6.31 for a shirt marked 25% off the regular price. What was the regular price of the shirt?

Problem 2.7 John bought a coat which usually sells for $98.00 at 25% off. What did he pay for the coat?

Problem 2.8 A coffee mug marked $15.00 was on sale for $12.00. What is the rate of discount?

Problem 2.9 A salesman who works on a commission basis earns 18% of his sales. How much was his commission on a $480 sale?

Problem 2.10 A clerk at the stationary store receives $12\frac{1}{2}\%$ commission on all merchandise sold. If she received $52 in commission last week, what were her sales for the week?

Problem 2.11 When gold sold for $16 an ounce, Johnny found $6 worth of gold in his claim. Gold presently sells for $328 an ounce. How many dollars is Johnny's amount of gold worth today?

Problem 2.12 The ratio of girls to boys participating in intramural volleyball at Ashland Middle School is 7 to 4. There are 42 girls in the program. What is the total number of participants?

Problem 2.13 If 5 bananas are worth the same as 3 apples then how many bananas are worth the same as 15 apples?

Problem 2.14 An agent receives a commission of 6% of the selling price of a house. The rest of the proceeds go to the owner of the house. If the agent sells a house for $135,000, what is her commission? How much does the house owner receive?

Problem 2.15 The ratio of the length of Mary's cat to the length of Amy's cat is 5:7. Mary's cat is 100 cm long. How much longer is Amy's cat than Mary's cat?

Problem 2.16 Thomas mixed 3 pints of red paint with 4 pints of blue paint to make a new color. He now wants to use 27 pints of red paint and some blue paint to make the same color. How many pints of blue paint will he need?

Problem 2.17 The ratio of boys to girls at the baseball game is $5 : 2$. There are 28 girls. How many more boys are there than girls?

Problem 2.18 Cathy and Jimmy looked for seashells at Newport Beach. For every 9 seashells Cathy found, Jimmy found 7. Cathy found 54 seashells. How many fewer seashells did Jimmy find than Cathy?

Problem 2.19 There are 2 packs of crayons available for every 5 students at Amy's art class. How many students can share 18 packs of crayons?

Problem 2.20 The ratio of boys to girls in Math Zoom Academy is 5:3. If there are 24 more boys than girls in Math Zoom Academy, then how many girls are in Math Zoom Academy?

2.3 Exercises - Two: Percentage Problems with More Depth

In this section we look at examples involving percentage with more depth than in the previous section. Sometimes, a problem involves more than one percentage, with respect to different base values.

Example 2.12

If 20% of a number is 30, what is 60% of the same number?

Solution

In this problem, the base value is the number itself. We need to find the number first. If this number is x, then

$$20\% \times x = 30,$$

thus

$$x = \frac{30}{20\%} = 150.$$

Now we can find the 60%:

$$60\% \times 150 = 90.$$

So the answer is 90.

Alternatively, if you recognize that 60% is three times the 20% quantities, you may get the answer immediately: $30 \times 3 = 90$.

Example 2.13

The tag price of a floral painting is $189.95. What is the net price if double discounts of 20% and 10% are applied?

Solution

It is generally incorrect to add the two percentage values to get a 30% discount. To calculate the correct price, we apply the discounts one by one:

$$189.95 \times (1 - 20\%) \times (1 - 10\%) = 189.95 \times 0.8 \times 0.9 = 136.764.$$

Therefore the net price is 136.76 after rounding.

Example 2.14

Connor gets 80% on a 10-problem test, 95% on a 20-problem test and 90% on a 30 problem test. If the three tests are combined into one 60-problem test, what percentage is his overall score?

Solution

To calculate the percentage of the overall score, we first find out the exact number of problems he answered correctly. On the first test, the number of correct answers is

$$80\% \times 10 = 0.80 \times 10 = 8;$$

on the second test, the number of correct answer is

$$95\% \times 20 = 0.95 \times 20 = 19;$$

on the third test, the number of correct answer is

$$90\% \times 30 = 27.$$

Therefore the total number of the questions he correctly answered is

$$8 + 19 + 27 = 54.$$

To calculate the percentage, we divide 54 by 60:

$$54 \div 60 = 0.90 = 90\%.$$

Therefore the overall score is 90%.

Problem 2.21 Sally is playing basketball. After Sally takes 20 shots, she has made 55% of her shots. She takes 5 more shots and she raises her percentage to 56%. How many of the last 5 shots did she make?

Problem 2.22 Michelle has a dozen oranges and a dozen pears. Assume all the oranges are the same size and all the pears are the same size. Michelle uses her juicer to extract 8 ounces of pear juice from 3 pears and 8 ounces of orange juice from 2 oranges. She makes a pear-orange juice blend from an equal number of pears and oranges. What percent of the blend is pear juice?

Problem 2.23 The number of students going on a field trip changed from 24 to 36. What was the percentage increase in the number of students going on a field trip? Later, the number of students going changed back from 36 to 24. What was the percentage decrease in the number? Give both answers as a percentage rounded to the nearest whole number.

Problem 2.24 Sixty percent of the people on a subway train are seated. As some people prefer standing, only 75% of the seats on the subway are filled. If there are 12 empty seats, how many people are on the train?

Problem 2.25 Carson bought five notebooks from the College Bookstore at a cost of $2.50 each. His brother Derick liked the notebooks and went to the bookstore the following day and also bought 5 notebooks. The bookstore had a 20%-off sale that day. How much did Derick save compared to Carson on the purchase of the five notebooks?

Problem 2.26 A collector offers to buy the 1967 year of the sheep stamp sheet for 2000% of its face value. Bridget has one of the sheets with 12 stamps that had an original face value of 25 cents per stamp. How much would Bridget receive if she sold it to the collector?

Problem 2.27 The ratio of llamas to ostriches in the Math Zoom Academy petting zoo is 4 : 7. If there are total of 44 llamas and ostriches in the petting zoo, how many of the them are llamas?

Problem 2.28 Henry reads 160 pages of a book per day. After 5 days, Henry has $\frac{3}{5}$ of the book remaining. How many pages does the book have?

Problem 2.29 In a far-off land three fish can be traded for two loaves of bread and one loaf of bread can be traded for six ears of corn. How many ears of corn are worth the same as one fish?

Problem 2.30 Jenny starts with a full jar of jellybeans. Each day, she eats 20% of the jellybeans that were originally in the jar. At the end of the second day, 36 jellybeans remain. How many jellybeans were in the jar originally?

Problem 2.31 Two-thirds of the monkeys in a cage are seated in three-fourths of the spots. The rest of the monkeys are standing. If there are 6 empty spots, how many monkeys are in the cage?

Problem 2.32 A shopper buys a $100 coat on sale for 20% off. An additional $5 is taken off the sale price by using a discount coupon. A sales tax of 8% is paid on the final selling price. What is the total amount the shopper pays for the coat?

Problem 2.33 The table below gives the percent of students in each grade at school A and school B.

	K	1	2	3	4	5	6
A	21%	12%	11%	15%	13%	17%	11%
B	18%	11%	16%	11%	13%	14%	17%

School A has 100 students, and school B has 200 students. If the two schools combined, what percent of the students are in grade 6?

Problem 2.34 Calvin bought four Avengers movie tickets for his friends at a cost of $12.50 each. His friend Mark wanted to watch the movie with his family as well and went to buy the same amount of tickets the following day. The theater had a 20%-off sale that day. How much did Mark save comparing to Karl on the purchase of four movie tickets?

Problem 2.35 In the popular TV show "Who Wants to be a Millionaire", contestants earn certain amount of money based on the number of questions they answer correctly. The dollar values of each question are shown in the following table (where k = 1000).

Question	1	2	3	4	5	6	7	8
Value	100	200	300	500	1k	2k	4k	8k
Question	9	10	11	12	13	14	15	
Value	16K	32K	64K	125K	250K	500K	1000K	

Between which two questions is the percentage increase of the value the smallest?

Problem 2.36 A middle school has 780 students, some of which go to Math Olympiad classes. Among those who attend Math Olympiad classes, $\frac{8}{17}$ are in 6th grade, and $\frac{9}{23}$ are in 7th grade. How many students do NOT attend Math Olympiad classes?

Problem 2.37 Jim is paid an 8% commission on the first $800 of weekly sales, and a 14% commission on any sales past $800. If Jim's sales were $1300, what was his commission?

Problem 2.38 Linda receives a weekly salary of $120 plus a commission of 5% on all sales above $500 per week. During three weeks Linda's total sales were $1540, $1235, and $1040. What was her total paycheck for the three weeks?

Problem 2.39 The table shows some of the results of a survey. What percentage of the males surveyed read the newspaper?

	Read	Don't Read	Total
Male	?	26	?
Female	58	?	96
Total	136	64	200

Problem 2.40 Rita has 36 marbles, 20 of which are red and 16 of which are white. Rose has 27 marbles, all of them either red or white. Suppose Rita and Rose have the same ratio of red to white marbles. How many more white marbles does Rita have than Rose?

2.4   Exercises - Three: Challenging Percentage Problems

This section contains challenging problems involving percentage. We shall go over a few examples first.

Example 2.15

Samantha's mathematics class had a test with 75 problems: 10 arithmetic, 30 algebra, and 35 geometry problems. Her friend Tori answered 70% of the arithmetic, 40% of the algebra and 60% of the geometry problems correctly, and she did not pass the test because she got less than 60% of the problems right. How many more questions would Tori have needed to answer correctly to earn a 60% passing grade?

Solution

To answer the question, we shall find out how many questions Tori answered correctly, and how many correct questions are needed to pass the test. We then can subtract the numbers to find out how many more questions Tori needed to get correct to pass.

It is straight forward to calculate the number of questions she already answered correctly:

$$70\% \times 10 + 40\% \times 30 + 60\% \times 35 = 7 + 12 + 21 = 40.$$

To pass the test, Tori needed to get 60% of the questions correct. Since there are 75 total questions,

$$60\% \times 75 = 45$$

correct questions are needed to pass. Thus, Tori needed to have answered

$$45 - 40 = 5$$

more questions to pass the test.

Example 2.16

Vitamin tablets are packed in three different sized bottles: small (S), medium (M) and large (L). The medium size costs 50% more than the small size and contains 20% fewer tablets than the large size. The large size contains twice as many tablets as the small size and costs 30% more than the medium size. Rank the three sizes from best to worst buy.

Solution

The question asks to find out which size is the best buy and which is the worst. This ranking is best obtained by calculating the unit price for each tablet.

We are not given any information about the actual prices and the actual amount of the vitamin in each packaging size. However, since all the relations between quantities are given in terms of percentage, we may assume a value for one of the quantities, and express everything else based on that value. We may treat the prices and the package size separately.

In terms of the price, we can assume the price of the small package is 100 cents (the unit of the price does not matter, we may assume it is 100 gold coins with no impact to the solution).

The medium package costs 50% more than the small one, so the price of medium package is
$$(100\% + 50\%) \times 100 = 1.50 \times 100 = 150$$
cents.

The large package costs 30% more than the medium one, so the price of the large package is
$$(100\% + 30\%) \times 150 = 1.30 \times 150 = 195$$
cents.

Now that we have the prices determined, let us work on the amounts of vitamins each packaging contains. Again, assume the small package contains 100 tablets.

The large package contains twice as much tablets as the small one, so the large package contains
$$100 \times 2 = 200$$
tablets.

The medium package contains 20% fewer tablets than the large one, so the medium package contains
$$(100\% - 20\%) \times 200 = 0.80 \times 200 = 160$$
tablets.

By now we have the prices and numbers of tablets for each packaging, we can calculate the unit prices per tablet. For clarity we display the data in the following table.

Size	Price	Num. Tablets	Unit Price
Small	100	100	$100/100 = 1$
Medium	150	160	$150/160 = 0.9375$
Large	195	200	$195/200 = 0.975$

From the table we see that the medium size package has the best unit price, thus it is the best buy. The large size one is the second best, and small size one is the worst buy.

Example 2.17

John is on a business trip with his friends Eric, Nick and Oscar. Oscar lost all his money, so the friends wanted to help him. Each gave Oscar the same amount of money. However, Eric gave Oscar 20% of his own money, John gave Oscar 25% of his money, and Nick gave Oscar $\frac{1}{3}$ of his money. What percent of the group's money does Oscar now have?

Solution

Don't we all wish we had friends like these?

The problem does not say how much money each of the friends gave to Oscar. Since all the amounts are described in percentage or fractions, and the question also asks for percentage, we may assume each friend gave Oscar $100. Oscar has received $100 from each of the three friends, so he has $300 now. To answer the question, we need to figure out the total amount of money they have, and it is necessary to find out the amount of money each of the friends had before giving $100 to Oscar.

Eric gave Oscar 20% of his money. This means $100 is 20% of Eric's money, so Eric had $100 \div 20\% = 500$ dollars to begin with.

John gave Oscar 25% of his money. Thus John had $100 \div 25\% = 400$ dollars before giving the Oscar.

Nick gave Oscar $\frac{1}{3}$ of his money. Therefore Nick had $100 \div \frac{1}{3} = 300$ dollars at the beginning.

Putting all these amounts together, the total amount of money they have is $500 + 400 + 300 = 1200$ dollars.

Oscar has 300 dollars now, and this is $\frac{300}{1200} = \frac{1}{4} = 25\%$ of the group's money.

Problem 2.41 Three bags of jelly beans contain 26, 28, and 30 beans. The bags consist of respectively 50%, 25%, and 20% yellow jelly beans. All three bags of beans are dumped into one bowl. What percent of all beans are yellow jelly beans? Round your answer to the nearest percent.

Problem 2.42 Andy had no money, so his Granny Smith gave him 36% of her money. Now Granny Smith has $80 left, and Andy has $2 more than Elberta. How many dollars does Elberta have?

Problem 2.43 Two 600 ml pitchers contain orange juice. One pitcher is 30% full and the other pitcher is 40% full. Water is added to fill each pitcher completely, then both pitchers are poured into one large container. What percent of the mixture in the large container is orange juice?

Problem 2.44 At a party there are only single women and married men with their wives. 40% of the women are single. What percentage of the people in the room are married men?

Problem 2.45 A and B are two identical cups. A is full with salt water containing 2% salt and B is half full with salt water containing 0.8% salt. Suppose we pour half of the salt water from cup A to cup B so cup B is now full of salt water. What percentage of salt is the salt water in cup B?

Problem 2.46 Business is a little slow at Lou's Fine Shoes, so Lou decides to have a sale. On Friday, Lou increases all of Thursday's prices by 10%. Over the weekend, Lou advertises the sale: "Ten percent off the list price. Sale starts Monday." How much does one pair of shoes cost on Monday that cost $40 on Thursday?

Problem 2.47 A merchant offers a large group of items at 30% off. Later, the merchant takes 20% off these sale prices and claims that the final price of these items is 50% off the original price. What is the true total discount?

Problem 2.48 A grocery store sells eggs in three sizes: small (S), medium (M) and large (L). The medium size costs 50% more than the small size and contains 20% fewer eggs than the large size. The large size contains twice as many eggs as the small size and costs 30% more than the medium size. Rank the three sizes from best to worst buy.

Problem 2.49 A bag contains 3 blue, 4 red and 3 yellow marbles. How many blue marbles must be added to the bag for it to contain 75% blue marbles?

Problem 2.50 Phil Lanthropist won a great deal of money in a contest. He gave 20% of his winnings to his parents, gave 25% of the remaining money to his children, and gave the remaining $900,000 to his favorite charity. What was the total number of dollars that Phil won?

Problem 2.51 In the fish tank at Albert's house, $\dfrac{1}{4}$ of the fish are red and the number of black fish is $\dfrac{3}{5}$ of the number of red fish. There are 24 additional fish that are all spotted. How many red fish are there?

Problem 2.52 The sales tax rate in Orange County is 8%. During a sale at an outlet in Orange County, the price of a suit is discounted 40% from its $190.00 price. Two clerks, Jimmy and Tony, calculate the bill independently. Jimmy first adds 8% sales tax to the price, and then subtracts 40% from this total. Tony first subtracts 40% of the price from the original price, and then adds 8% sales tax to the discounted price. What is Jimmy's total minus Tony's total?

Problem 2.53 A fruit salad consists of crabapples, cranberries, black berries, and black cherries. If there are twice as many cranberries as crabapples and three times as many black berries as black cherries and four times as many black cherries as cranberries and the fruit salad has 280 total fruits, then how many black cherries does it have?

Problem 2.54 Jong-Zhi took a math test that had 12 arithmetic questions, 15 algebra questions and 18 geometry questions. She got 75% of the arithmetic questions correct and 60% of the algebra questions correct. How many of the geometry questions must she get correct to get a passing grade of 75%?

Problem 2.55 A $480 TV was put on sale for 30% off. It wasn't sold so the price was lowered an additional percent off the sale price making the new price $285.60. What percentage was the second discount?

Problem 2.56 Chris's pay went from $20/hr to $25/hr after his first evaluation. After his second evaluation his pay was raised to $33/hr. What is the difference between the second raise as a percent and the first raise as as percent?

Problem 2.57 John, Edward, and Dan did a fundraiser for the math club at school and raised a total of $370. They divided the $370 into three parts such that the second part is $\frac{1}{4}$ of the third part and the ratio between the first and the third part is 3 : 5. Find the value of each part.

Problem 2.58 Sally baked 5 dozen cookies on Saturday afternoon. She gave 60% of the cookies to her neighbors at the neighborhood barbecue. On Sunday, she took 75% of the remaining cookies to the church social. On Monday night, she and her family ate 50% of the remaining cookies while watching football. What percent of the 5 dozen cookies remain?

Problem 2.59 In her history class, Marie averaged 90% correct on five 10 question quizzes, got 96% correct on a 50 question midterm exam, and 75% correct on an 80 question final exam. What is the percentage of correct answers she provided if the total points for correct answers on all quizzes and exams are combined?

Problem 2.60 The students in Miss Einstein's class took a math test. Two-thirds of the class passed and the other third failed. The ratio of boys to girls in the class is 2 to 1. All of the girls passed the exam. What percentage of boys failed the exam?

3. Chickens and Rabbits

Samantha Wincher enjoys participating in her school plays. This year, they are putting on the Lion King. To represent the animals, the four-legged animals are played by two people who have a cover over their heads that is decorated like the animal. For the two-legged animals, one person wears a costume. The main characters are played by one person, so that it's easier for them. Samantha's role is one of the main characters, but she also has a side role of the minor animals. She crouches behind her friend and can only see other people's legs while under the cover. Knowing that there are thirty animals in total in this scene, and that the only animals present at this moment are antelope and ostriches. Since all the legs are human legs, and she can't see above that, Samantha counts how many legs there are in total and comes up with one hundred legs. She wonders how many of each animal there are. How can she find out without throwing off the costume cover?

The Chicken-Rabbit problem is a classical problem in ancient Chinese text. The math concepts used to solve this kind of problems only includes the basic arithmetic operations of addition, subtraction, multiplication, and division, but it also requires creative problem solving skills. Some people may say that you have to know algebra—systems of linear equations involving two variables. In fact, algebra is not necessary. Instead of using algebra, we shall learn some creative problem solving strategies, and those strategies can produce much faster solutions.

The concepts introduced in this chapter directly correspond to Common Core Math Standards as shown in the following table.

6th Grade	6.RP.3, 6.EE.7
7th Grade	7.NS.3, 7.EE.3, 7.EE.4
8th Grade	8.EE.7, 8.EE.8

In addition to the standards above, problems and concepts in this section will help strengthen understanding of the following domains.

6th Grade	6.RP, 6.NS, 6.EE
7th Grade	7.RP, 7.NS, 7.EE
8th Grade	8.EE

3.1   Key Concepts

The typical chicken-rabbit problem is the following. There are chickens and rabbits in the farm. We are given the number of heads and the number of legs, and are required to find out the number of each type of animals.

Two possible directions can be followed for solving this kind of problems. One direction is to use the basic arithmetic operations you learned in elementary school: addition, subtraction, multiplication, and division. Plus, a good use of your critical thinking skills! The creative problem solving methods developed hereby will demonstrate why this kind of problems became the classics in the ancient Chinese wisdom. You will see how it works in the solutions below.

For people who know algebra, you may easily set up a linear equation or a system of two linear equations, and solve those equations or systems and get the answers. Those are perfectly good solutions. You may say this is not challenging at all! However, without using equations, the methods are more interesting and usually faster, and less error-prone.

3.2 Exercises - One: Basic Chicken & Rabbit Problems

> **Example 3.1**
>
> There are some chickens and some rabbits on a farm. Suppose there are 50 heads and 140 feet in total among the chickens and rabbits, how many chickens are there? How many rabbits?

There are many different approaches to solving the standard chicken and rabbit type of problem. Here we examine three different types of solutions.

Solution 1

First of all, each animal has one head. A chicken has two feet, and a rabbit has four feet. There are 50 heads, so there are 50 animals.

Now assume all 50 animals are chickens, there should be total 100 feet. But there are $140 - 100 = 40$ additional feet. These 40 feet came from the rabbits which were wrongfully assumed to be chickens. Each rabbit has 2 more feet than the wrongfully assumed chicken, so we need to count in two additional feet for each rabbit. Now $40 \div 2 = 20$, which is the number of rabbits. The rest is the number of chickens, which is 30.

In conclusion, there are 20 rabbits, and 30 chickens.

Remark

> Note: For clarity of explanation, we described the problem solving process in separate steps. Once you get used to this process, you will be able to go through this thinking process very quickly in your head, so you can solve the problem with a matter of seconds.

Solution 2

Step 1. We may use a little bit more imagination this time. Imagine all the animals are trained with special skills for a circus performance. When the trainer blows a whistle, each chicken stands with one foot, and each rabbit stands up with two hind legs. All the animals lift half the number of feet up to the air. On the ground now, there should be $140 \div 2 = 70$ feet.

Step 2. Now each chicken has exactly one foot on the ground. However many chickens we have, we should count exactly same number of feet. But each rabbit would have two feet on the ground. Therefore, the extra number of feet on the ground 70, in comparison to the number of heads given, which is 50, would be the number of rabbits: $70 - 50 = 20$.

Step 3. We now know that there are 20 rabbits, so the number of chicken is $50 - 20 = 30$.

In conclusion, there are 20 rabbits, and 30 chickens.

Remark

> Again, for clarity of explanation, the problem solving process is split into 3 steps. Once you get used to this process, you will be able to go through this thinking process very quickly in your head, so you can solve the problem with a matter of seconds.

Solution 3

Now let's use the algebraic method. Let x be the number of chickens, and y be the number of rabbits. There are 50 animals in total, so

$$x + y = 50.$$

Each chicken has 2 feet, while each rabbit has 4. Since there are 140 feet in total, we have

$$2 \times x + 4 \times y = 140.$$

This gives us the system of equations

$$\begin{cases} x + y &= 50, \\ 2x + 4y &= 140. \end{cases}$$

To solve this system of two equations, we use substitution method. The first equation can be be written as

$$x = 50 - y.$$

Substituting this into the second equation we have

$$2 \times (50 - y) + 4y = 140,$$

so distributing and combining like terms gives us

$$2y = 40.$$

Hence we can solve for y to get

$$y = \frac{40}{2} = 20.$$

Plugging this back in we also get

$$x = 50 - y = 50 - 20 = 30.$$

Thus there are 20 chickens and 30 rabbits.

Remark

Variations of the given solutions: For each solution presented above, you could alter a little bit and create a new solution. For example, in Solution 1, you could assume all the animals are rabbits instead.

In Solution 2, you could pretend the wings of the chickens were feet, and make them the same number of feet as the rabbits, which means 4 feet for each animal, for a total of $4 \times 50 = 200$ feet. Now you wonder what the 60 extra feet are . . . (hint: we called the wings "feet")

In Solution 3, you could use only one variable, and directly set up the one-variable linear equation, and solve for the variable.

Example 3.2

Mary is baking a lot of cookies for a bake sale at Samantha's school. To bake the cookies she needs a total of 60 eggs. The eggs come in small cartons containing 6 eggs or large cartons containing 12 eggs. If Mary buys a total of 7 cartons, how many small cartons and how many large cartons does she buy?

Solution

If Mary bought 7 small cartons of eggs, she would have a total of

$$7 \times 6 = 42$$

eggs, which is

$$60 - 42 = 18$$

less than she needs. Hence some of the cartons must be large cartons. Since a large carton contains

$$12 - 6 = 6$$

extra eggs, if Mary replaces

$$18 \div 6 = 3$$

of the small cartons with large cartons she will have the correct number of eggs. Therefore Mary buys

$$7 - 3 = 4$$

small cartons of eggs and 3 large cartons of eggs.

Problem 3.1 There are some chickens and some rabbits on a farm. Suppose there are 45 heads and 128 feet in total among the chickens and rabbits, how many of the animals are chickens? How many are rabbits?

Problem 3.2 In a math competition, there are 25 questions. Each correct answer earns 6 points. One point is taken away for each incorrectly answered or unanswered question. Jenny received 101 points. How many questions did she answer correctly?

Problem 3.3 Two trucks dump dirt of 400 cubic meters. Truck A carries 7 cubic meters per load. Truck B carries 4 cubic meters per load. The dirt is removed after 70 loads. How many loads are carried by truck A?

Problem 3.4 The price of a pack of colored pencils is $19 and the price of a pack of regular pencils is $11. The math teacher bought 16 packs of pencils for a total of $280. How many packs of each type did he buy?

Problem 3.5 Sarah counts her chickens and rabbits, and there are 16 heads and 44 feet. How many of each type are there?

Problem 3.6 In a farm there are goats and ducks. The total number of heads is 100, and the total number of legs is 316. How many animals of each type are there?

Problem 3.7 Sixty vehicles (cars and motorcycles) are parked in a parking lot. Totally there are 190 wheels. Given that a car has 4 wheels and a motorcycle has 2 wheels, how many cars and motorcycles each are in the parking lot?

Problem 3.8 Teachers and students from the Areteem Summer Camp visited the museum. They bought a total of 99 tickets for 218 dollars. If each teacher ticket costs 4 dollars, and each student ticket costs 2 dollars, how many teachers and students were there respectively?

Problem 3.9 The counselor brought his 51 students to the lake to go rowing, 6 people for each big boat and 4 people for each small boat. They rented 11 boats to fit everyone with no empty seats. How many big and small boats each did they rent?

Problem 3.10 Each set of chess is played by 2 students, and each set of Chinese checkers is played by 6 students. A total of 26 sets of chess and Chinese checkers are played by exactly 120 students in a school event. How many sets of each game are there?

Problem 3.11 Use 400 matches to make triangles and pentagons. Totally 88 triangles and pentagons are made with no matches left over. How many of each shape are made?

Problem 3.12 There are 20 questions in a math competition. Five points are given to each correct answer, and -3 points are for each incorrect answer or unanswered question. Jeff received 60 points in the competition. How many questions did he answer correctly?

Problem 3.13 Tiffany scored 29 points in her school's playoff basketball game. She made a combination of 2-point shots and 3-point shots during the game. If she made a total of 11 shots, how many 3-point shots did she make?

Problem 3.14 There are 48 tables in a restaurant. Small tables can seat 2 people, and big tables can seat 5 people. They can accommodate a maximum number of 159 people. How many small tables and how many big tables are there?

Problem 3.15 In Bob's Cycle Shop, workers received the delivery of an order Bob placed for the single seat bicycles and tricycles. There are a total of 90 seats and 215 wheels, plus all the other necessary parts and accessories. How many bicycles and how many tricycles can they assemble?

Problem 3.16 The company Green Pilots is organizing a company picnic at the beach. They want to save energy by carpooling to the beach site. They are able to fit 450 people into 80 vehicles that are either 5-seat sedans or 7-seat minivans. How many sedans and how many minivans do they need to use to take everyone to the picnic?

Problem 3.17 The Math Zoom Camp allows students to form four-person teams or six-person teams to work together on projects. If there are 42 teams formed with the 200 total students in the camp, how many four-person teams and six-person teams are formed?

Problem 3.18 17 people went to a farm. There were goats to feed and chickens to feed, and each person fed exactly one animal. It costs $1.50 to feed a chicken and $2.00 to feed a goat. In total, the people spent $32.50. How many of each type of animal did they feed?

Problem 3.19 A large family of 20 people goes to a restaurant. They each order either pizza or salad. The pizza costs $9.00 and salad costs $3.00. In all the family spends $138.00. How many pizzas and how many salads did the family order?

Problem 3.20 A pet owner has cats and birds. There are 25 pets in total and all combined the pets have a total of 90 legs. How many of each are there?

3.3   Exercises - Two: Chicken & Rabbit Problems with More Depth

Example 3.3

There are some chickens and some rabbits on a farm. Suppose there are 50 heads and there are 20 more rabbit feet than chicken feet on the farm. How many chickens and how many rabbits are there? ♣

Solution 1

If all 50 animals are rabbits, there are

$$50 \times 4 = 200$$

rabbit feet and 0 chicken feet, so there are 200 more rabbit feet than chicken feet. If we replace a rabbit with a chicken, we remove 4 rabbit feet and add 2 chicken feet, so the difference between rabbit feet and chicken feet decreases by

$$4 + 2 = 6.$$

Since we know there are 20 more rabbit feet than chicken feet, which is

$$200 - 20 = 180,$$

smaller than if all the animals were rabbits, we must replace

$$180 \div 6 = 30$$

rabbits with chickens. Hence there are

$$50 - 30 = 20$$

rabbits and 30 chickens on the farm.

It is also possible to first assume there are 50 chickens and proceed similarly. We also give an algebraic solution.

Solution 2

Let x be the number of chickens and y be the number of rabbits. There are 50 heads, so

$$x + y = 50.$$

We know there are 20 more rabbit feet than chicken feet, so

$$4 \times x = 2 \times y + 20.$$

Rearranging we have the system of equations

$$\begin{cases} x + y & = & 50, \\ 4x - 2y & = & 20. \end{cases}$$

Doubling the first equation we have

$$2x + 2y = 100$$

and adding this to the second equation we have

$$6x = 120.$$

Hence

$$x = \frac{120}{6} = 20,$$

so substituting we have

$$20 + y = 50,$$

so

$$y = 50 - 20 = 30.$$

Hence there are 20 rabbits and 30 chickens.

Example 3.4

Connor is helping his biology teacher get ready for the class's next laboratory. The teacher needs 10 liters of a 68% alcohol solution for the students to disinfect their tools. Connor makes this solution by mixing a 80% alcohol solution with a 50% alcohol solution. How many liters of each solution did Connor use?

Solution

Since Connor creates 10 liters of 68% alcohol solution, he uses a total of

$$68\% \times 10 = 6.8$$

liters of pure alcohol. If he used only the 50% solution, he would only have

$$50\% \times 10 = 5$$

liters of pure alcohol, which is

$$6.8 - 5 = 1.8$$

liters less than he actually needs. Every liter of 80% solution he uses instead of a liter of 50% solution has

$$80 - 50 = 30$$

---

percent more alcohol (so contains 0.3 extra liters of alcohol). Hence, Connor used

$$1.8 \div 30\% = 1.8 \div 0.3 = 6$$

liters of the 80% alcohol solution. The other

$$10 - 4 = 6$$

liters are therefore the 50% alcohol solution.

Remark

> Other solutions, including an algebraic solution with a system of equations, are possible. See how many alternate solutions you can find!

Problem 3.21 Sasha takes a mathematics competition. There are a total of 20 problems. For each correct answer, competitors receive 5 points. For each wrong answer, they instead get 1 point taken away. Sasha has 64 points in total. How many problems did she answer correctly?

Problem 3.22 Aria has 16 coins that are nickels and dimes. The total value is $1.20. How many nickels and dimes does Aria have?

Problem 3.23 There are 100 birds and cats. The birds have 80 more legs than the cats. How many birds and how many cats are there?

Problem 3.24 How many gallons of a 25% alcohol solution must be mixed with a 50% alcohol solution to make 30 gallons of a 40% alcohol solution?

Problem 3.25 A group of 68 people rent 24 motorcycles of two kinds at a racetrack. The first kind has a capacity of 2 people and costs $40 per motorcycle. The second has a capacity of 3 people and costs $30 per motorcycle. The 68 people exactly fill all vehicles. What is the total cost in renting the 24 motorcycles?

Problem 3.26 There are many ducks and sheep in a farm. If we count the heads, there are total of 80 heads. If we count the legs, there are 56 more legs from sheep than from ducks. How many ducks and how many sheep are there in the farm.

Problem 3.27 Bella goes shopping at the marketplace for shawls and belts. The shawls she likes each cost $12. The belts she likes each cost $14. Bella has exactly enough money to buy a certain number of shawls. If she buys belts instead, she has exactly enough money to buy 3 fewer belts. How much money did Bella bring with her to the market?

Problem 3.28 Debbie is mixing orange juice concentrate for her restaurant. The first juice concentrate is 64% real orange juice. The second is only 48% real orange juice. How many ounces of 48% real orange juice should she use to make 1600 ounces of 58% real juice?

Problem 3.29 The Math Club collected donations from 40 people who live in either City A or B. Each person from City A contributed $5, and each person from City B contributed $8. In total, $5 more was collected from City A than from City B. How many people are there in each city?

Problem 3.30 A butcher has some hamburger meat that is 4% fat and some hamburger meat that is 20% fat. How much of each type will he need to make 120 pounds of hamburger meat which is 10% fat?

Problem 3.31 Hank has a bottle of diluted syrup that is 60% maple syrup and a bottle of pure syrup that is 100% maple syrup in his restaurant. How many ounces of each should he mix in order to make 100 ounces of 85% maple syrup?

Problem 3.32 How many gallons of 60% antifreeze should be mixed with 40% antifreeze to make 80 gallons of 45% antifreeze?

Problem 3.33 It requires either 45 small trucks or 36 big trucks to transport a batch of steel blocks. Given that each big truck can load 4 more tons of steel blocks than each small truck. How many tons of steel blocks are in a batch?

Problem 3.34 In a farm the total number of chickens and rabbits is 100. If the number of chicken feet is 80 more than the number of rabbit feet, how many chickens and rabbits are there respectively?

Problem 3.35 Four basketballs and five volleyballs cost 185 dollars in total. If a basketball costs 8 dollars more than a volleyball, what is the cost of one basketball?

Problem 3.36 A turtle has 4 legs and a crane has 2 legs. There are totally 100 heads of turtles and cranes, and there are 20 more crane legs than turtle legs. How many of each animal are there?

Problem 3.37 On a good day, Chris the Squirrel picks 20 hazelnuts. On a rainy day he only picks 12 hazelnuts. During a few consecutive days he picked a total of 120 hazelnuts with an average of 15 per day. How many days were rainy?

Problem 3.38 In a farm there are 6 times as many rabbits as chickens, and the total number of feet of the chickens and rabbits is 390. What is the number of each type of the animals?

Problem 3.39 The capacity of a big container is 4 gallons, and that of a small container is 2 gallons. 50 containers are filled with water, and there are totally 20 more gallons of water in the big containers than the small containers. How many big and small containers are there respectively?

Problem 3.40 Morgan needed 70 sticks for a project at school. Each stick is either 3 inches or 5 inches and the total length of all the sticks combined is 270 inches. How many 3 inch sticks and 5 inch sticks are there?

3.4 Exercises - Three: Challenging Chicken & Rabbit Problems

With extra information, the same methods can be used to solve more complicated chicken and rabbit problems.

> **Example 3.5**
>
> Suppose there are chicken, rabbits, and sheep on a farm. There are 70 heads in total and 220 feet. If there are the same number of rabbits and sheep, how many chickens, rabbits, and sheep are on the farm?

Solution

There are an equal number of rabbits and sheep, so pair up one rabbit with one sheep. This pair has 2 heads and 8 legs. If there are no chickens, there are

$$70 \div 2 = 35$$

such pairs, for a total of

$$35 \times 8 = 280$$

feet. This is a total of

$$280 - 220 = 60$$

extra feet. Replacing a rabbit/sheep pair with 2 chickens reduces the number of feet by

$$8 - 4 = 4.$$

Hence there are

$$60 \div 4 = 15$$

pairs of chickens and thus

$$35 - 15 = 20$$

rabbit/sheep pairs. Thus there are 30 chickens, 20 rabbits, and 20 sheep on the farm.

Example 3.6

Connor is classifying bugs for a biology project. He has 15 bugs in total. Some are spiders, with 8 legs. Some are houseflies, with 6 legs and 1 pair of wings. The remaining are dragonflies, with 6 legs and 2 pairs of wings. If there are 98 legs and 17 pairs of wings, how many of each type of bug does Connor have?

Solution

If we first assume all 15 bugs are houseflies, then there will be

$$15 \times 6 = 90$$

legs. There are in fact 98 legs, which is

$$98 - 90 = 8$$

more than if we all bugs were houseflies. Each dragonfly has the same number of legs but each spider has

$$8 - 6 = 2$$

---

extra legs, so there must be

$$8 \div 2 = 4$$

spiders in total. The number of houseflies and dragonflies is thus

$$15 - 4 = 11.$$

Again, assume these 11 bugs are all houseflies. There will be 11 pairs of wings, which is

$$17 - 11 = 6$$

less than the actual amount. Each dragonfly has an extra pair of wings, so there must be 6 dragonflies. Hence the number of houseflies is

$$11 - 6 = 5.$$

To summarize there are 4 spiders, 5 houseflies, and 6 dragonflies.

Remark

> An algebraic approach will work here as well, but now we will have a system of equations with 3 equations and 3 unknowns.

Problem 3.41 Some chickens and rabbits have a total of 100 feet. If each chicken was exchanged for a rabbit, and each rabbit was exchanged for a chicken, there would be a total of 86 feet. How many chickens are there? How many rabbits?

Problem 3.42 100 mice eat 100 cakes. If each big mouse eats 3 cakes, and 3 baby mice eat 1 cake, how many big mice and baby mice are there?

Problem 3.43 A spider has 8 legs. A firefly has 6 legs and 2 pairs of wings. A cicada has 6 legs and 1 pair of wings. There are a total of 16 bugs of the three types. There are 110 legs in total. There are 14 pairs of wings in total. How many of each kind of bug are there?

Problem 3.44 The school purchases 3 different sizes of projectors, total of 47. The large size costs $700, the medium costs $300, and the small costs $200. The total cost of the projectors is $21200, and there are twice as many medium projectors than the small. How many large projectors does the school purchase?

Problem 3.45 Tony's mom took out $380 from the bank. There are $2, $5, and $10 bills and total of 80. The number of $5 bills and $10 bills are the same. How many bills of each type are there?

Problem 3.46 Cindy collects 20 insects for her biology class, all of which are spiders, dragonflies, and cicadas. (Note that a spider has 8 legs and no wings, a dragonfly has 6 legs and 4 wings, and a cicada has 6 legs and 2 wings.) She counts 138 legs and 36 wings altogether. How many insects are there in each kind?

Problem 3.47 A candy shop sold three flavors of candies, cherry, strawberry, and watermelon, in the morning. The prices are $20/kg, $25/kg, and $30/kg, respectively. The shop sold a total of 100 kg and received $2570. It is known that the total sale of cherry and watermelon flavor candies combined is $1970. How many kilograms of watermelon flavor candies were sold?

Problem 3.48 A crab has 10 legs. A mantis has 6 legs and 1 pair of wings. A dragonfly has 6 legs and 2 pairs of wings. There are a total of 37 of the three types. There are 250 legs in total. There are 52 pairs of wings in total. How many of each kind are there?

Problem 3.49 100 monks eat 100 steamed buns. If each senior monk eats 4 steamed buns, and 4 junior monks eat 1 steamed bun, how many senior monks and junior monks are there?

Problem 3.50 A large bottle can hold 4 liters of oil, while every two small bottle can hold 1 liter of oil. A store has 100 liters of oil and the oil exactly fills up 60 bottles. How many of each kind of bottle does the store have?

Problem 3.51 Charles and David are fast typists. Charles can type 12 more words per minute than David. Charles started typing, and 2 minutes later David also started typing, and they both stopped after 3 more minutes. Given that they typed 780 words altogether, how many words did each of them type?

Problem 3.52 In the warehouse there were 3 times as much apples as bananas at the beginning. Suppose 250 pounds of bananas and 600 pounds of apples were sold every day, and a few days later the bananas were sold out and 750 pounds of apples were left. How many pounds of each fruit were there originally?

Problem 3.53 Jack went hiking. His uphill speed was 3 miles per hour and downhill speed was 5 miles per hour, and he hiked a total of 6 hours including both uphill and downhill, with total distance 23 miles. How many hours did he spent uphill and downhill respectively?

Problem 3.54 Lily spent $490 to buy 80 color pencils for her art class, including red, green. and blue colors. The red pencils cost $2 each, the green ones cost $5 each, and the blue ones cost $10 each. Suppose she bought the same number of green and blue pencils. How many of each type of pencils did she buy?

Problem 3.55 A spider has 8 legs and no wings. A dragonfly has 6 legs and 2 pairs of wings. A cicada has 6 legs and one pair of wings. There are a total of 18 bugs of these types, with 118 legs and 20 pairs of wings. How many dragonflies are there?

Problem 3.56 Some friends rent some boats. If 4 people get in each boat, 8 people will not fit in the boats. If 5 people get in each boat, 6 people will not fit in the boats. How many friends are there? How many boats?

Problem 3.57 A grocer mixed grape juice, which costs \$2.25 per gallon, with cranberry juice, which costs \$1.75 per gallon. How many gallons of each should be used to make 200 gallons of a cranberry/grape juice mix that costs \$2.10 per gallon?

Problem 3.58 The owner of the Fancy Food Shoppe wishes to mix cashews selling at \$8.00 per kilogram and pecans selling at \$7.00 per kilogram. How much of each kind of nut should be mixed to get 8 kg worth \$7.25 per kilogram?

Problem 3.59 A convenience store owner wishes to mix together raisins and roasted peanuts to produce a high energy snack for hikers. The raisins sell for \$3.50 per kilogram and the nuts sell for \$4.75 per kilogram. How many kilograms of each should be mixed together to obtain 20 kg of this snack with a price of \$4.00 per kilogram?

Problem 3.60 A meat distributor paid \$2.50 per pound for hamburger meat and \$4.50 per pound for ground sirloin. How many pounds of each did he use to make 100 pounds of meat mixture that will cost \$3.24 per pound?

4. Motion—Speed, Time, and Distance

Getting around and going to places are what we do on a daily basis. It's one of the fundamental human activities in life to go from one place to another. Like Samantha, students walk to school every morning, and walk home after school. Like Mr. Winchers, parents drive to work. Grandparents take the bus to the market. Families fly to a faraway place for vacation.

The study of movement of an object is classified as Motion. In this chapter, we study some basic Motion problems. To describe the motion of an object, there are three elements involved: Speed, Distance, Time.

Speed indicates how fast an object travels in a unit of time measurement. For example, we use miles or kilometers per hour to describe how fast a motorized vehicle moves or an airplane flies. Generally speaking, driving a car is faster than walking, and taking an airplane is faster than driving a car. We may use a smaller unit such as meters per second to describe how fast an insect moves.

Time is another key factor in motion. We say how many hours to travel from Los Angeles to San Diego, and we say how many minutes it takes for you to walk to school. The unit of measure used is based on the understanding of the surrounding world.

Distance is a key factor as well when we talk about motion. We say how many miles or kilometers from a city to another, or from home to school or to work. We use meters or feet to describe how long the path is in your backyard.

Speed, time, and distance are the three important factors in motion; we will see real life examples that is relevant to motion in this chapter.

The fundamental equation in motion problems is

$$\text{rate} \times \text{time} = \text{distance}$$

or

$$rt = d.$$

Here the term "rate" has the same meaning as "speed". Focusing on rate or time, we may also express the equation as

$$r = \frac{d}{t} \qquad \text{or} \qquad t = \frac{d}{r}.$$

These equations are the fundamental formulas for the motion problems. There could be very interesting and sophisticated logic built in the enriched motion problems for real life scenarios. You might say that motion problems are easy because you have already known the formula and practiced with some word problems from school textbooks. However, there are much more to explore on this topic.

The concepts introduced in this chapter directly correspond to Common Core Math Standards as shown in the following table.

6$^\text{th}$ Grade	6.RP.2, 6.RP.3, 6.EE.7
7$^\text{th}$ Grade	7.RP.2, 7.NS.3, 7.EE.3, 7.EE.4
8$^\text{th}$ Grade	8.EE.7

In addition to the standards above, problems and concepts in this section will help strengthen understanding of the following domains.

6$^\text{th}$ Grade	6.RP, 6.NS, 6.EE
7$^\text{th}$ Grade	7.RP, 7.NS, 7.EE
8$^\text{th}$ Grade	8.EE

4.1 Key Concepts

> **Average Speed**
>
> The average speed over a certain traveled distance or over a period of time is defined as the total distance divided by the total time, regardless of the instantaneous speed at any particular instant of time during the trip.
>
> $$\text{Average speed} = \frac{\text{Total distance}}{\text{Total time}}.$$

Remark

Most of the time it is wrong to simply add up the speeds of different segments and divide by the number of segments.

For problems related to the average speed, it is always beneficial to focus on the total distance and total time.

> **Example 4.1**
>
> Mr. Winchers drives a car from home to a client site in Wilmington early morning at the speed of 72 miles per hour. On his way back home from the client site in Wilmington, the traffic is getting bad; he drives the car back home at a reduced speed of 48 miles per hour. What is his average speed for the round trip?

Solution

As mentioned earlier, we cannot simply add 72 and 48 and get 120, and divide 120 by 2 to get 60 miles per hour, which is the wrong answer. To get the correct average speed, we compute the total distance and total time. However, we are not given the distance in this problem. To continue, we can use a variable d to indicate the one-way distance in miles. So the time Mr. Winchers takes to reach the destination is $\dfrac{d}{72}$ hours, and the return trip takes $\dfrac{d}{48}$ hours. For the round trip, the total distance is $2d$ miles and the total time spent is $\dfrac{d}{72} + \dfrac{d}{48} = \dfrac{5d}{144}$ hours. To find the average speed, we calculate the total distance divided by the total time:

$$2d \div \frac{5d}{144} = 288 \div 5 = 57.6$$

in miles per hour.

Remark

Note that the variable d eventually cancels out and does not affect the final answer. In fact, you could assume a distance that is easy to calculate, such as 144 miles, which is a common multiple of 72 and 48, and calculate an average speed based on this distance: the total distance would be 288 miles, and it takes Mr. Winchers 2 hours to reach the destination and 3 hours for the return trip, totally 5 hours, and the average speed would be $288 \div 5 = 57.6$ miles.

Practice: Please use a few different assumed values for the distance and verify that the answers are same.

4.1.1 Motion in Flowing Water

In this section, we explore the scenario where an object, such as a boat, moves in flowing water. Both the speed of the object and the speed of the water flow are considered in the analysis.

Suppose R is the speed of the object (such as a boat) in still water, and r is the speed of the water flow.

- If the movement is downstream, the actual speed is $R + r$;
- If the movement is upstream, the actual speed is $R - r$.

Remark

The same methods can be used if the movement is in the air with wind blowing.

Example 4.2

Connor and his friends went to river tubing in the Math Zoom Summer Camp. Connor sat still on the tube and enjoyed the beautiful scenes on the river bank, with no effort made to pilot the tube. He left the starting point at 3PM and arrived ♣ at the end at 4:15PM for the 2 and a quater mile distance. How fast did the water flow on the river? Use miles per hour for the speed.

Solution

Since Connor was sitting on the tube effortless, so his speed would be same as that of the water flow. The total distance is $2\frac{1}{4}$ miles; The total time is 75 minutes, which is $1\frac{15}{60}$ hours. To find the average speed, we calculate the total distance divided by the total time: So the average speed $= 2\frac{1}{4} \div 1\frac{15}{60} = \frac{9}{4} \div \frac{75}{60} = \frac{9}{4} \times \frac{60}{75} = \frac{9}{4} \times \frac{4}{5} = \frac{9}{5} = 1.8$ miles per hour.

Example 4.3

On the same tubing trip, Derek used two paddling sticks to forge forward. He ended up arriving at the ending point at 3:45PM. What was Derek's paddling speed?

Solution

Derek's overall speed can also be calculated with the basic motion formula by dividing the total distance by the total time. The total distance is $2\frac{1}{4}$ miles. His total time was $\frac{45}{60}$ hours. Derek's overall speed $= 2\frac{1}{4} \div \frac{45}{60} = \frac{9}{4} \div \frac{45}{60} = \frac{9}{4} \times \frac{60}{45} = \frac{9}{4} \times \frac{4}{3} = \frac{9}{3} = 3$ Since the water speed is 1.8 miles per hours, Derek's paddling speed would be the difference between his overall speed and the water flow speed. Derek's paddling speed $= 3 - 1.8 = 1.2$ in miles per hour.

4.1.2 Relative Speed and Frame of Reference

Relative speed is the speed of an object in reference to another moving object, or we can say it's the speed between two objects in motion, without regards to any fixed reference frame.

In the example above, Derek's speed in reference to the river bank would be his overall speed. However, in reference to the water, his speed would be his paddling speed.

Relative speed

The relative speed is the speed between two objects in motion without regards to any fixed reference.

Frame of Reference

The frame of reference is a set of criteria or stated values in relation to which measurements or judgments can be made. In the study of motion, it refers to a system of geometric axes in relation to which measurements of position or motion can be made.

4.1.3 **Meet Up, Farewell, and Catch Up**

In this section we explore objects moving toward each other, away from each other, or along the same direction.

Example 4.4

Imagine Samantha is chasing a running dog. If Samantha and the dog are both running at 5 miles per hour in reference to a fixed object like a house, then their relative speed is 0. It means from the Samantha's point of view if she focuses on the dog, she feels no change of the distance between herself and the dog. In this case, the relative speed is the difference between the two objects' speed.

In a different scenario that the dog is sitting still, Samantha is running toward the dog, the relative speed is Samantha's running speed.

Or else, if Samantha and the dog were running toward each other, Samantha would feel they are getting closer at a much faster pace. This is because the relative speed is the sum of Samantha's and the dog's running speed.

Remark

When two objects are moving in opposite directions their speeds should be added to find the relative speed. When two objects are moving in the same direction the relative speed is the difference of their speeds.

4.1.4 Problem Solving Strategies for Motion Problems

> **Problem Solving Strategies for Motion Problems**
>
> - For Average Speed problems: focus on the total distance and total time.
> - For Motion in Flowing Water problems:
>   - The difference between the downstream speed and upstream speed is twice the water flow speed.
>   - Sometimes it is better to switch the frame of reference to the water. That is, pretend you are floating with the water, and from your point of view the water does not move. As a result, the object's upstream and downstream speeds are the same, and objects on the bank (such as piers) move at the flow speed in the opposite direction.
> - To find the relative speed:
>   - When two objects are moving in opposite directions, find the sum of their speeds.
>   - When two objects are moving in the same direction, find the difference between their speeds.
> - Draw diagrams or charts to keep track of the information given in the problem.

The next few examples demonstrate the problem solving strategies for motion problems.

> **Example 4.5**
>
> Samantha needs to run a mile for less than 10 minutes for the PE class. What is her least average speed in terms of Miles per Hour and Meters per Second? (1 mile = 1609 meters)

Solution

There are 60 minutes in one hour. If Samantha has to run 1 mile in 10 minutes, then she will have to run 6 miles in one hour, her least average speed = 60 miles per hour.

Since 1 mile = 1609 meters, and 1 hour = 3600 seconds, $1609 \div 3600 = 0.45$ meters per second.

Example 4.6

The Winchers family is taking a road trip from Los Angeles to Phoenix Arizona. The distance is about 360 miles. If they drove at a constant speed of 60 miles per hour for the first 150 miles, and drove at a constant speed of 70 miles per hour for the remaining 210 miles, how long did it take them to get there?

Solution

Since the term "average" is involved, we look at the total distance. This is given in the problem: 360 miles. In the first 150 miles, the speed is 50 miles per hour, thus it took them $150 \div 60 = 2.5$ hours. For the remaining distance 210 miles, they drove 70 miles per hour, so it took them $210 \div 70 = 3$ hours. The total time is $2.5 + 3 = 5.5$ hours. Let's use the formula, the average speed = $360 \div 5.5 = 65.45$ miles per hour.

Example 4.7

Let's make the question a little bit different. The Winchers family is taking a road trip from Los Angeles to Phoenix Arizona. The distance is about 360 miles; they planned to get there in 6 hours. If they drove at a constant speed of 50 miles per hour for the first 120 miles, at what constant speed do they need to drive for the remaining 240 miles in order to get there in time?

Solution

Since the term "average" is involved, we look at the total distance. This is given in the problem: 360 miles. In the first 120 miles, the speed is 50 miles per hour, thus it took $120 \div 50 = 2.4$ hours. In order to get there in 6 hours, they need to drive the remaining 240 miles in $6 - 2.4 = 3.6$ hours, and the speed would be $240 \div 3.6 = 200 \div 3 = 66.7$ miles per hour.

Example 4.8

Mr. and Mrs. Winchers were rowing a boat upstream in the river. At exactly 1pm, Mrs. Winchers's hat fell into the water, but they didn't find out until sometime later. They immediately turned around and rowed downstream, and caught up with the hat 6 miles from the point the hat fell. Assuming the water flow speed is 4 miles per hour. (1) What is the exact time when they found the hat? (2) Assuming they rowed at a constant speed relative to the water, what was the time when they realized the hat fell and turned around?

Solution

At first glance, there seem to be missing information. We don't know their rowing speed. However, since the hat was floating with the river, it took 1.5 hours for the hat to move 6 miles downstream, and that was how long it took for them to find the hat. Therefore they found the hat at 2:30pm. For the second question, we use the hat as the frame of reference, then the water was still and the two people rowed the boat at a constant speed, so they took the same amount of time away and back. Therefore they turned around at exactly 1:45pm.

Remark

In this problem, we don't know the rowing speed, and it actually doesn't matter what the speed was.

Example 4.9

The Hare and the Tortoise are having a race. The Hare's speed is 20 miles per hour, and the Tortoise's speed is 1 mile per hour. At the beginning, the Hare takes a nap and gave the Tortoise 5 miles head start. Then the Hare wakes up and runs after the Tortoise. How long does it take for the Hare to catch up with the Tortoise?

Solution

The relative speed between the Hare and the Tortoise is $20 - 1 = 19$ miles per hour, and the distance between them is 5 miles to begin with. It will take $5/19$ of an hour for the Hare to catch up.

Example 4.10

Adam lives in Town A, and Bob lives in Town B, and the two towns are 10 miles apart. At the same time, Adam starts walking from Town A to Town B at 3 miles per hour, and Bob starts walking from Town B to Town A at 2 miles per hour, and a bird also starts flying from Adam towards Bob. Once the bird meets Bob, it turns back towards Adam, and once it meets Adam it turns towards Bob again, and so on, until Adam and Bob meet. Assuming the bird flies at 15 miles per hour. What is the total distance that the bird flies?

Solution

The sum of the speeds of Adam and Bob is 5 miles per hour, so it takes them 2 hours to meet. The bird flies for 2 hours without stopping, and that is totally 30 miles.

4.2 Exercises - One: Basic Motion Problems

First of all, let's get very comfortable with using the basic formula of the speed, time and distance.

Problem 4.1 Suppose a train travels a distance of 120 miles in 3 hours. What is the average speed of the train?

Problem 4.2 Mary and Amy rollerblade at an average speed of 9 miles per hour for 3.5 hours, how far will they travel?

Problem 4.3 In a cross-country race, Tony drove his car for 707 kilometers (km) in 7 hours. What was his average speed?

Problem 4.4 A tennis ball is thrown a distance of 20 meters. What is its speed if it takes 0.5 seconds to cover the distance?

Problem 4.5 A bat is flying at a speed of 45 kilometers per hour. How much time does it take to travel a distance of 1,800 kilometers?

Problem 4.6 Adam takes a train to go visit a friend who lives in a city that is 360 kilometers away. The train left his home station at 8:35AM, and arrived at the destination station at 1:05PM. How fast has the train traveled measured by average speed?

Problem 4.7 Thomas and his family went on a road trip last week. They traveled 50 mph from Chicago, IL to Minneapolis, MN and 65 mph on the return trip. What was the average speed for the entire round trip?

Problem 4.8 Jimmy and his brother took a circular ride at an amusement park on averages of 30 miles per hour and took them $2\frac{1}{2}$ minutes. Roughly how big is the diameter of the circular track?

Problem 4.9 If I drive from Irvine to Fullerton at 60 miles per hour and then from Fullerton to Irvine at 40 miles per hour, what is my average speed for the whole journey?

Problem 4.10 Katie went hiking on a hill near her home. From the bottom of the hill, She went up to the top and then came down along the same trail, back to the spot she started. Assume her uphill speed was 3 miles per hour, and her downhill speed was 6 miles per hour. What is her average speed for the whole uphill-downhill trip?

Problem 4.11 Melisa drove for 3 hours at a rate of 50 miles per hour and for 2 hours at 60 miles per hour. What was her average speed for the whole journey?

Problem 4.12 Frank drives his car for a distance of 300 miles. For the first 135 miles, he drives at a constant speed of 45 miles per hour. At what constant speed does he drive for the remaining distance to average 50 miles per hour?

Problem 4.13 Suppose a truck travels in segments that are described in the table below:

Segment	Distance (miles)	Time (hours)
1	30	1
2	90	2
3	50	1

What is the average speed of the truck?

Problem 4.14 Joe and his family are planning to go to a national park which is 600 miles away from the home. How fast in miles per hour must they drive if they want to get there in 15 hours?

Problem 4.15 Ling goes mountain hiking in a park. She first walks uphill at a speed of 2.5 miles per hour, and she next walks downhill at a speed of 4 miles per hour. The round trip takes 3.9 hours. What is the distance for the round trip?

Problem 4.16 Stephanie begins walking at a pace of 4 km per hour from one end of the trail that is 34 km long. Bob begins at the other end of the trail at the same time, walking towards Stephanie at a pace of 6 km. How long will it take for them to pass each other?

Problem 4.17 Terry and Susan are entered in a 24-mile race. Susan's average rate is 4 miles per hour and Terry's average rate is 6 miles per hour. Both start at the same time. How far will Susan be away from the finish line when Terry crosses the line?

Problem 4.18 John's house and Mary's house are 14 miles apart. They start at noon to walk toward each other in order to go to a book fair together. John walks at a rate of 3 mph, and Mary walks at a rate of 4 mph. How many hours will it take them to meet?

Problem 4.19 Jasmine took a walk after dinner. She first walked 5 km in 1.5 hours, and then walked for 1 km in 0.5 hour in the same direction. What is her average speed for the whole journey?

Problem 4.20 Two friends leave the same place at the same time traveling in the same direction. One travels at a speed of 55 mph and the other travels at a rate of 65 mph. After 2 hours, how many miles will they be away from each other?

4.3   Exercises - Two: Motion Problems with More Depth

Problem 4.21 Mike drives his car for a round trip between LA and San Diego. He drives at 70 miles per hour to get from LA to San Diego. At what speed should he drive back, if his average speed for the round trip is 60 miles per hour?

Problem 4.22 Alice leaves site A toward site B at the same time Bob leaves site B toward A. Alice drives at 40 miles per hour, and Bob drives at 60 miles per hour. After they pass by each other, Alice drives 4.5 additional hours to arrive at B. How far is it between A and B?

Problem 4.23 Linda goes mountain biking in a park. She first bikes on flat road at a speed of 12 miles per hour, then goes uphill at a speed of 9 miles per hour, and she next bikes downhill at a speed of 18 miles per hour, along the same trail as the uphill trip. Finally she goes back home along the same flat road she traveled earlier. The round trip takes 4 hours. What's the distance for the round trip?

Problem 4.24 As Emily is riding her bicycle on a long straight road, she spots Emerson skating in the same direction 1/2 mile in front of her. After she passes him, she can see him in her rear view mirror until he is 1/2 mile behind her. Emily rides at a constant rate of 12 miles per hour, and Emerson skates at a constant rate of 8 miles per hour. For how many minutes can Emily see Emerson?

Problem 4.25 Two trains, 121 meters and 99 meters in length respectively, are moving in opposite directions, one at the rate of 40 km/h and the other at the rate of 32 km/h. In what time will they be completely clear of each other from the moment they meet?

Problem 4.26 Two trains, 121 meters and 99 meters in length respectively, are moving in the same direction, one at the rate of 40 km/h and the other at the rate of 32 km/h. How long would it take to be completely clear of one another, if the faster train has just met up with the back of the slower train.

Problem 4.27 A train, 110 meters in length, travels at 60km/h. In what time will it pass a man who is walking at 6km/h (i) against it; (ii) in the same direction?

Problem 4.28 A train moving 25 km/h takes 18 seconds to pass a platform. Next, it takes 12 seconds to pass a man walking at 5 km/h in the opposite direction. Find the length of the train and of the platform.

Problem 4.29 It takes 6 hours for an airplane to fly a round trip. If the speed of the airplane is 1500 km per hour on the departure trip, and 1200 km per hour on the return trip. What is the one-way distance ?

Problem 4.30 Starting at the same time, Tom and Jerry walk toward each other. Tom walks from site A to B at 5 miles per hour. Jerry walks from B to A. After they meet, Jerry walks an additional 10 miles to arrive at A, and Tom spends additional 1.6 hours to walk and arrive at B. What is the rate at which Jerry walks?

Problem 4.31 Starting at the same time, Heather and Brenda drive their cars from site A toward site B. Heather drives at 52 km per hour, and Brenda drives at 40 km per hour. After 6 hours of driving, Heather passes a truck traveling in the opposite direction. One hour later, Brenda passes the same truck still traveling in the opposite direction. At what speed does the truck travel?

Problem 4.32 David and Ray hike a mountain trail in Crystal Cove. David starts out on the trail at a pace of 4 kilometers per hour. One hour later, Ray starts out on the same trail at 6 kilometer per hour. How long will it take Ray to catch up to David?

Problem 4.33 Sam begins walking at a pace of 4 km per hour from one end of a trail that is 34 km long. Ashley begins one hour later at the other end of the trail, walking towards Sam at a pace of 6 km per hour. How long will it take for them to pass each other?

Problem 4.34 Tom drives his car for a round trip between place A and place B. He drives at 40 km per hour to get from A to B. At what speed should he drive back from B to A, if his average speed for the round trip is 48 km per hour?

Problem 4.35 A boat has a rip-hole on the bottom while 20 miles away from the shore. The water comes in at a rate of 1.5 tons every minute, and the boat would sink after 70 tons of water came in. How fast must the boat go in order to reach the shore before sinking?

Problem 4.36 One day, Bob rode his bike to school. When school is off, he forgot his bike and walked home instead. He spent a total of 50 minutes on the road for the round trip. If he walked for both directions, he would have spent a total of 70 minutes. How much would be the total time if he rode his bike for both directions?

Problem 4.37 It takes 40 minutes for Dave to walk from home to school. It takes 15 minutes if he rides a bike instead. One day, he first rides a bike for 9 minutes before the bike breaks. He then walks the remaining distance to school. How much total time does Dave spend getting to school?

Problem 4.38 Starting at the same time, a bus and a truck start traveling toward each other. After 18 hours the two vehicles meet. The bus travels at 50 miles per hour. The truck travels at 42 miles per hour, but stops for a 1 hour break after every 3 hours of travel. What is the distance between the two starting locations?

Problem 4.39 It takes a pigeon 2 hours to fly with the wind between two houses, and 3 hours against the wind. If the wind blows at a speed of 2.5 miles per hour, at what speed would the pigeon travel in a windless day?

Problem 4.40 Brian and David run along a circular track, starting from the same point, going opposite directions. They meet after 36 seconds. Assume that David runs the whole circle in 90 seconds. How long does it take Brian to run the whole circle?

4.4   Exercises - Three: Challenging Motion Problems

Problem 4.41 An ant crawls along the sides of an equilateral triangle. It starts at one vertex and crawls at a rate per minute of 50 cm, 20 cm and 40 cm, respectively, on each of the three sides of the triangle. What is the ant's average speed as it travels around the triangle?

Problem 4.42 A road consists of uphill, flat and downhill sections with that order. The distances of the three sections are in ratio $1 : 2 : 3$ with a total distance of 20 miles. The times JoAnn spends on the three sections are in ratio $4 : 5 : 6$. She walks at a speed of 2.5 miles per hour uphill. What is the total time she spends on the road?

Problem 4.43 A bridge consists of three sections of equal length: an uphill section, a flat section and a downhill section. At what average speed does a cyclist ride his bicycle if he travels the three sections at a speed of 4 meters per second, 6 meters per second and 8 meters per second respectively?

Problem 4.44 At 6 AM, bus station A starts to dispatch buses to station B, and station B starts to dispatch buses to station A. They each dispatch one bus to the other station every 8 minutes. The one-way trip takes 45 minutes. One passenger gets on the bus at station A at 6:16 AM. How many buses coming from station B will the passenger see en route?

Problem 4.45 Sam walks up a hill. After every 30 minutes of walking he takes 10 minutes to rest. When he walks down the hill, he instead rests for 5 minutes after every 30 minutes of walking. Sam walks downhill 1.5 times faster than he walks uphill. If he spends 3 hours and 50 minutes traveling up the hill, how much time does he spend traveling down the hill?

Problem 4.46 Omar walks up a hill. After every 40 minutes of walking uphill he takes 10 minutes to rest. Downhill he rests for 5 minutes after every 40 minutes of walking. Omar walks downhill at a speed 2 times as fast as that he walks uphill. If he spends 2 hours traveling down the hill, how much time does he spend traveling up the hill?

Problem 4.47 It takes a ship 6 hours to travel downstream between two piers, and 8 hours upstream. If the water flows at a speed of 2.5 miles per hour, at what speed would the ship travel in still water?

Problem 4.48 George goes to school by riding his bike to the bus station, taking the bus, and then walking to his classroom. The ratio of the three distances is 2:8:1. His biking speed is 10 mph. The bus travels at a speed of 50 mph. His walking speed is 2 mph. What is his average speed in his way to school?

Problem 4.49 Cindy rides her bike from home to school at a speed that is 120 meters per minute faster than if she walks, and the time she spends is 3/5 less than if she walks. How fast does Cindy walk from home to school?

Problem 4.50 Joe and JoAnn walk toward each other from two locations that are 36 miles apart. If Joe departed 2 hours earlier, they would meet 2.5 hours after JoAnn departed. If JoAnn departed 2 hours earlier, they would meet 3 hours after Joe departed. Find the respective speed at which each walks.

Problem 4.51 Sami and Rajan practice running together. If Sami starts to run after Rajan runs for 10 meters, then it will take Sami 5 seconds to catch up with Rajan. If Sami starts to run after Rajan runs for 2 seconds, then it will take Sami 4 seconds to catch up with Rajan. How fast can each person run?

Problem 4.52 It takes 25 seconds for a train to pass completely pass through a tunnel which measures 250 meters long. It takes 23 seconds for the train to completely pass through another tunnel which measures 210 meters long. How long does it take the train to pass an approaching train which is 320 meters long and at the speed of 18 m/s?

Problem 4.53 Ming and Ping both take a walk. The distance that Ming walks is 1/5 less than the distance that Ping walks. Also, the time Ping spends is 1/8 more than the time Ming spends. What is the ratio of their respective speed?

Problem 4.54 Tyler and Hannah start to walk on the same direction from the same place. Tyler walks at 5 miles per hour. Hannah walks at 1 mile per hour for the first hour, 2 miles per hour for the second hour. Hannah increases her speed by 1 mile per hour after each hour. How long does it take for Hannah to catch up with Tyler?

Problem 4.55 A boat takes 3 days to travel from town A to town B, but it takes 4 days to travel from town B to town A. If a motor-less raft is left alone in the water by town A, how long will it take for the raft to float to town B?

Problem 4.56 Starting at the same time, Cathy and David drive two cars toward each other from the two ends, call them A and B, of the same road. Cathy drives 1.2 times faster than David. When they pass by each other, they are 8 miles away from the halfway point between A and B. Find the total length of the road.

Problem 4.57 Alice, Bob, and Cindy drive their cars separately from site A to site B simultaneously. Alice drives at 60 mph and Bob drives at 48 mph. Alice passes a car from the opposite direction after 6 hours of driving. One hour later, Bob pass the same car still traveling in the opposite direction. One more hour later, Cindy also passes the same car. Find the speed at which Cindy drives her car.

Problem 4.58 Amy and Amanda walk on a circular track. Amy starts from spot A, and Amanda starts from spot B. One walks clockwise and the other counter-clockwise. After 6 minutes, they meet. Four minutes later, Amy arrives at spot B. After 8 more minutes of walking, they meet again. How many minutes does it take for each to walk one full circle?

Problem 4.59 A railroad bridge measures 1000 meters long. A train passes the bridge. It takes 120 seconds from the time the train enters the bridge to the time the whole train gets off the bridge. There are 80 seconds during which time the whole train is on the bridge. Find both the speed and the length of the train.

Problem 4.60 A hunting dog chases a hare 21 meters ahead. The dog runs in a series of jumps, with each jump being 3 meters long. Each jump for the hare is 2.1 meters. If the dog jumps three times for every four times the hare jumps, how much farther can the hare travel before the dog catches it?

5. Work Related Problems

Samantha loves baking, so she volunteered to help prepare and bake for a school party. One of the foods she has chosen to make is pretzels. She has baked these before, and her whole family enjoyed them. One thing would help her, though: another person! Samantha's best friends, Jo, Ella, and Charlie all want to help.

The day before the party, Samantha wants to calculate how much time it would take to prepare and bake the pretzels if she had a helper, so that she can plan out the rest of her day. Since three of her friends have volunteered, she wonders: how much time would it take to bake if one of my friends came over? What about two? Or three? She knows how long it would take for her to bake them by herself; how can she calculate how long it will take if one, two, or three of her friends assisted her?

Work related problems involve scenarios of certain amount of work shared by individuals, where each individual work at a different rate. For example, 3 people paint a wall, 2 people mow a lawn, 5 people peel potatoes, etc.

The concepts introduced in this chapter directly correspond to Common Core Math Standards as shown in the following table.

6th Grade	6.RP.2, 6.RP.3, 6.EE.7
7th Grade	7.RP.2, 7.NS.3, 7.EE.3, 7.EE.4
8th Grade	8.EE.7

In addition to the standards above, problems and concepts in this section will help strengthen understanding of the following domains.

6th Grade	6.RP, 6.NS, 6.EE
7th Grade	7.RP, 7.NS, 7.EE
8th Grade	8.EE

5.1   Key Concepts

Work-related Problems

The typical work-related problem involves the following components: total amount of work to be done, the speed at which the work is done by a person or a group of people, and the time spent doing the work.

The relationship among the components follow the equation

$$\text{Work} = \text{Working speed} \times \text{Time}.$$

Remark

> Motion Problems can also be treated as a type of work-related problems, if we consider the distance to be traveled as the total amount of work to be done, and the moving rate as the working speed.

If the total amount of work is considered as a whole (usually denoted with the number 1), and the number of unit time periods taken to get the job done is X, then the amount of work done in each unit time period is $\frac{1}{X}$. (Based on the scenarios in the problems, the unit time period can be any unit of time: seconds, minutes, hours, days, weeks, months, years, etc.)

Fractions and their operations, including addition, subtraction, multiplication, and division, are essential in solving work-related problems.

Example 5.1 Job Sharing

Samantha invites Jo to help with her baking. Samantha usually prepares the pretzels for 40 minutes before putting them to the oven. Today she has Jo for help, so they worked together to prepare the pretzels. Being less experienced, Jo would have taken 60 minutes to prepare the pretzels all by herself. Working together, how long would it take for them to prepare the pretzels? In addition, it takes 15 minutes for the prepared pretzels to be baked. What is the total time needed for Samantha and Jo to prepare and bake the pretzels together?

Solution

To answer these questions, it is not necessary to know how many pretzels they are making. We simply use the quantity 1 to represent the whole task of preparation before baking. It takes Samantha 40 minutes to complete, therefore Samantha prepares $\frac{1}{40}$ of the task in each minute. For Jo, she does $\frac{1}{60}$ of the task per minute. Working together, they can complete

$$\frac{1}{40} + \frac{1}{60} = \frac{3}{120} + \frac{2}{120} = \frac{5}{120} = \frac{1}{24}$$

of the task. So working together, they can complete the preparation in 24 minutes. Because it takes an additional 15 minutes to bake the pretzels, the total time needed is $24 + 15 = 39$ minutes.

As it turned out, Samantha didn't need Ella or Charlie to help with the baking. So Ella and Charlie also brought some drinks and refreshments to the school party, and they had a very good time. Everyone liked the pretzels Samantha and Jo baked. They were so yummy!

Example 5.2 Filling or Draining Water Tanks

The community swimming pool held 18000 gallons of water when it was full. One day, Samantha went swimming but found that the pool was being drained for cleaning. If the draining pump drained the water at the rate of 12 gallons per minute, how long would it take for the pool to be completely drained?
The answer to the question above is

$$18000 \div 12 = 1500 \text{ minutes,}$$

which equals $1500 \div 60 = 25$ hours. Obviously Samantha could not swim that day. After the pool was drained, another water pump was used to fill the pool with fresh water. If the rate of filling the pool was 60 gallons per minute, how long would it take for the pool to be refilled with fresh water?
The answer to this new question is

$$18000 \div 60 = 300 \text{ minutes,}$$

which equals 5 hours.

Example 5.3 Filling or Draining Water Tanks (continued)

While anticipating the time that she would be able to swim again in the pool, Samantha had a strange thought: what if the people forgot to turn off the draining pump while pumping in fresh water? The pool would gain 60 gallons of water per minute while losing 12 gallons per minute, and the net effect would be gaining $60 - 12 = 48$ gallons of water per minute, and the pool would be filled in

$$18000 \div (60 - 12) = 18000 \div 48 = 375 \text{ minutes,}$$

which equals $6\frac{1}{4}$ hours.

Filling or Draining Water Tanks

In general, starting with an empty tank,

Time to fill $=$ Total volume $\div$ (incoming rate $-$ outgoing rate).

In the case the outgoing rate is higher than the incoming rate, the water would be drained. Starting with a full tank,

Time to drain $=$ Total volume $\div$ (outgoing rate $-$ incoming rate).

5.2 Exercises - One: Basic Work Problems

To solve work related problems, it's important to identify the components and clarify their relationships. Use diagrams to visualize the components to make it easier to find the ratio of work speed between the people who share the job.

Problem 5.1 Eldridge can split a cord of wood in 4 days and his father can do it in 3 days. How long would it take them if they worked together?

Problem 5.2 Using a ride-on lawn mower, Abby can mow the lawn in 2 hours. Her sister Carla takes 3 hours using an older mower. How long will it take them if they work together?

Problem 5.3 One drain pipe can empty a swimming pool in 6 hours. Another pipe takes 3 hours. If both pipes are used simultaneously to drain the pool, how long does it take the drain the pool?

Problem 5.4 Tom and Jerry can finish organizing the books at school's library together in 5 hours. If Tom do it alone, it will take him 8 hours. How long would it take Jerry to finish the same task alone?

Problem 5.5 Stan can load the truck in 40 minutes. If I help him, it takes us 15 minutes. How long will it take me alone?

Problem 5.6 Suppose Chris can paint the entire house in fourteen hours, and Bill can do it in ten hours. How long would it take the two painters working together to paint the house?

Problem 5.7 Two mechanics in the maintenance department were working on Daniel's car. One can complete the maintenance service in 5 hours, but the other mechanic, who is new, needs 8 hours. How long would it take the two mechanics working together to finish the service?

Problem 5.8 Billy and Tim can paint a fence in 4 hours together. It is known that Billy can paint the same fence alone in 6 hours. How long would it take Tim to paint the fence alone?

Problem 5.9 Kathy takes 3 hours to wash 300 dishes, and Andrew takes 2.5 hours to wash 300 dishes. How long will they take together to wash 1100 dishes?

Problem 5.10 Amy and her sister Clair's house has a 420 square foot lawn in the back. Amy can mow 120 square feet in 30 minutes. When Amy and Clair work together, they can finish the whole lawn in one hour. How many square feet per minute can Clair mow?

Problem 5.11 If 8 people can finish a task in thirteen days, how many days would it take to do the same job with 16 people?

Problem 5.12 It takes 1.5 hours for Jim to water the plants in a garden. Lily can water the same amount of plants in 2 hours. How long will it take Jim and Lily, work together to water the plants in the garden?

Problem 5.13 It takes pump A 6 hours to fill the pool, pump B 8 hours, and pump C 4.8 hours. How long would it take the three pumps together to fill the pool?

Problem 5.14 Nancy can row a boat across the river in 45 minutes, while Susan can do it in 35. If both of them sit in one boat and row together, how long will it take?

Problem 5.15 Employee A can complete a task in 3 hours. When working with Employee B, they can complete it in 2 hours. How long does it take for Employee B to finish the task if he/she works alone?

Problem 5.16 Niki always leaves her cell phone on. If her cell phone is on but she is not actually using it, the battery will last for 24 hours. If she is using it constantly, the battery will last for only 3 hours. Since the last recharge, her phone has been on 9 hours, and during that time she has used it for 60 minutes. If she doesn't talk any more (but leaves the phone on), how many more hours will the battery last?

Problem 5.17 There is a 10,000 liter swimming pool in Lance's community. The pool has two pipes: A and B. Pipe B delivers 1,000 liters water per hour. When pipe A and B are both on, the pool can be filled in 4 hours. How many liters per hour can pipe A deliver?

Problem 5.18 Bob, John, and Calvin can paint a wall alone in 2 hours, 2.5 hours, and 1.5 hours, respectively. How long does it take if all three of them work together?

Problem 5.19 Joe's mom can clean the kitchen in 45 minutes. If Joe helps his mother, they can clean it in 30 minutes. How long would it take Joe to clean it by himself?

Problem 5.20 Melisa can finish a project in 2 hours, while Terry works 1.5 times faster than Melisa. How long would it take them to finish together?

5.3 Exercises - Two: Work Problems with More Depth

Problem 5.21 Janelle cleans her aquarium by replacing $\frac{2}{3}$ of the water with new water, but that doesn't clean the aquarium to her satisfaction. She decides to repeat the process, again replacing $\frac{2}{3}$ of the water with new water. How many times will Janelle have to do this so that at least 95% of the water is new water?

Problem 5.22 It will take a Type A robot 6 min to weld a fender, but a Type B robot takes only $5\frac{1}{2}$ minutes. If the robots work together for 2 min, how long will it take the Type B robot to finish welding by itself? Express your answer as a mixed number.

Problem 5.23 Phil can paint the garage in 12 hours and Rick can do it in 10 hours. They work together for 3 hours. How long will it take Rick to finish the job alone?

Problem 5.24 Lincoln can do a job in 8 hours and Dave can do it in 6 hours. What part of the job can they do by working together for 2 hours? For x hours?

Problem 5.25 Andrew spends 5 hours to complete a quarter of a job. Charles spends 6 hours to complete half of the remaining part of the job. How much more time does it take for Andrew and Charles to work together to get the rest of the job done?

Problem 5.26 It takes Rianna 24 days to finish a job. For Helen, it takes 32 days. Rianna works on it for some days before Helen takes it over, and it takes a combined total of 26 days to get the job done. How many days does Rianna work on the job?

Problem 5.27 It takes Jacqueline 50 minutes to type a draft. For Virginia, it takes 30 minutes. Suppose after Jacqueline types for some time, Virginia types the rest of the draft, and it takes a combined total of 42 minutes. What fraction of the draft did Jacqueline type?

Problem 5.28 It takes Elizabeth 9 hours to complete a project. For Tiffany, it takes 12 hours. If they take turns, starting with Elizabeth, each working for one hour at a time, how much total time does it take for them to complete the project?

Problem 5.29 If Emily and Julia work together, they can finish a project in 6 days. It takes same amount of time for Emily to complete $\frac{1}{2}$ of the project as it takes for Julia to complete $\frac{1}{3}$ of the project. How long does it take for Emily alone to complete the project?

Problem 5.30 Carolyn reads a book. Initially, the number of pages that she had read and the number of pages that she has not read are in ratio 3 : 4. After she reads an additional 33 pages of the book, the ratio becomes 5 : 3. How many pages does the whole book have?

Problem 5.31 Adam can mow his entire yard in three hours. His sister, Brooke, can mow $\frac{3}{4}$ of the same yard in two hours. Using two identical mowers, what part of the yard can they mow in one hour working together?

Problem 5.32 One pipe can fill a swimming pool 1.5 times faster than a second pipe. If the gardener opens both pipes, they fill the pool in 5 hours. How long would it take to fill the pool if only the slower pipe is used? How about only the faster pipe?

Problem 5.33 Bill usually takes 50 minutes to groom the horses. After working for 10 minutes, he was joined by Ann and they finished the grooming in 15 minutes. How long would it have taken Ann working alone?

Problem 5.34 Mona can complete a task alone in 150 minutes. Sarah can finish the same task in 3 hours. They work together for 30 minutes, and then a new worker, Li, joins them, and they finish the task 30 minute later. How long would it take Li to finish the task alone?

Problem 5.35 If pump A is used alone, it takes 6 hours to fill the pool. Pump B takes 8 hours alone to fill the same pool. Uncle Sam wants to use three pumps: A, B and C to fill the pool in 2 hours. What should be the rate of pump C in order to accomplish Uncle Sam's goal?

Problem 5.36 Patty and Tracy can finish decorating a house for the holidays in 2.5 hours if they work together. Patty works twice as fast as Tracy. How long would it take to each of them if they work alone?

Problem 5.37 Peter can paint a wall in 40 minutes and John can paint the wall in 60 minutes. If they work together for 12 minutes, how much of the wall is left unpainted?

Problem 5.38 James, Patty, and Joseph can organize the new products in the warehouse in 2 hours. If James does the job alone, he can finish in 5 hours. If Patty does the job alone, she can finish it in 6 hours. How long will it take for Joseph to finish the job alone?

Problem 5.39 There is a drain and a hose in the pool. It is known that the hose can fill the pool in 21 hours, and the drain can empty the pool in 24 hours. How long does it take to fill the pool if the drain is open at the same time?

Problem 5.40 A senior worker and a new worker together produce a set of machines. The senior worker can produce 40 pieces per hour, and the new worker can produce 30 pieces per hour. When they finished, the new worker produced exactly 450 pieces. How many pieces did they produce in total?

5.4 Exercises - Three: Challenging Work Problems

Problem 5.41 When two teams A and B work together, it takes 18 days to get a job completed. After team A works for 3 days, and team B works for 4 days, only $\frac{1}{5}$ of the job is done. How long does it take for team A alone to complete the job? For team B alone?

Problem 5.42 Suppose if Ethan works for 5 days and Owen for 6 days, they finish a project. Alternatively, if Ethan works for 7 days and Owen for 2 days, they also finish the project. How long does it take for Ethan alone to complete the project? For Owen alone?

Problem 5.43 Students from the Key Club at Whitman High School wash cars from the two parking lots A and B. There are four times as many cars in lot A than in lot B. First, they wash the cars in lot A for half a day. Next, half of students continue, and half of students start to wash cars in lot B. The work in lot B is done at the end of the day. Unfortunately, there are still cars unwashed in lot A. If all the students worked together to finish the cars in lot A, how long would it take?

Problem 5.44 If Iris spends 3 days and Olivia spends 5 days on a project, $\frac{1}{2}$ of the work can be completed. If instead Iris spends 5 days and Olivia spends 3 days on the project, $\frac{1}{3}$ of the work can be completed. How long does it take to complete the whole project if Iris and Olivia work together?

Problem 5.45 A pool has an inlet pump and an outlet pump. If the pool is empty, and the inlet pump is open, it takes 5 hours to fill the pool with water. If the pool is full, and the outlet pump is opened, it takes 7 hours to empty the pool. Suppose after the inlet pump is open for 2 hours, both the pumps are opened. How much longer does it take for the pool to be half full of water?

Problem 5.46 It takes $\frac{1}{3}$ more time for Andy to plant one tree than for Nathan. If Andy and Nathan work together, then in the end Nathan plants 36 more trees than Andy does. How many trees are in total?

Problem 5.47 Anthony can cut a lawn in 2 hours, Mia can cut the same lawn in 3 hours, and Dandria can cut the same lawn in 2 hours. Anthony cuts the lawn for $\frac{1}{2}$ hour, and then Mia replaces Anthony and cuts the lawn for 1 hour herself. How many additional minutes will it take Dandria to finish cutting the lawn by herself?

Problem 5.48 A pool has two inlet pumps A and B. If pump A alone is open, it takes 12 hours to fill the pool with water. If pump B is open, it takes 18 hours to fill the pool with water. If the pool needs to be filled in 10 hours, what is the least amount of time both pumps need to be open?

Problem 5.49 There are two pumps A and B. They are used to fill water in two pools that, if full, hold equal amount of water. The ratio of the water-pumping rate for pump A and pump B is $7 : 5$. After $2\frac{1}{3}$ hours, the water in the two pools, if combined, can exactly fill one pool. Next pump A increases the water-pumping rate by 25%, pump B reduces the water-pumping rate by 30%. After pump A fills the pool, how much longer does it take for pump B to fill the second pool? Express your answer as a mixed number.

Problem 5.50 Brandon, Richard and Samuel are friends. They build a wall divider in a yard. Brandon and Richard work together, and they build $\frac{1}{3}$ of the divider in 5 days. Next, Richard and Samuel work together, and they build $\frac{1}{4}$ of the rest of the divider in 2 days. Last, Brandon and Samuel work together, they finish the rest in 5 days. How long does it take for Brandon alone to build the divider? For Richard alone? For Samuel alone?

Problem 5.51 A certain number of small parts need to be produced. 30 parts are scheduled to be produced after each day. After $\frac{1}{3}$ of the parts are produced, the rate of production increases by 10% thanks to improvement in efficiency. It takes 4 fewer days to produce all the parts than scheduled. How many parts are in total?

Problem 5.52 Peter painted $\frac{1}{3}$ of a room while Richard painted $\frac{2}{5}$ of the same room. It then took Peter 1 hour, 40 minutes to finish painting the remainder of the room by himself. In how many hours could Peter have painted the entire room by himself? Express your answer as a mixed number.

Problem 5.53 Adam and Bob are each assigned a task to paint a wall. The two walls are identical. At the beginning, Adam went to the wrong wall and painted 500 square feet on Bob's wall. At this moment Bob came and found Adam's mistake. Adam then returned to his own wall and Bob continued painting his wall. After a few days, Bob finished his task, but Adam is not yet done with his job. Now Bob decided to come and help Adam. Bob painted 1000 square feet on Adam's wall. Which person did more of the job?

Problem 5.54 A pool can be filled by pipe A in 3 hours, and pipe B in 5 hours. When the pool is full, it can be drained by pipe C in 4 hours. Suppose the pool is empty and all three pipes are open, how long will it take to fill up the pool?

Problem 5.55 Robot A can produce 48 pieces of machinery per hour, and robot B can produce 36 pieces per hour. After they worked together for 8 hours, there were 64 pieces marked as defective when the company tested them. How many non-defective pieces they can produce together in one hour?

Problem 5.56 Company A and B plan to manufacture a number of TVs together. After Company A worked for 6 days, it had finished $\frac{1}{4}$ of the TVs. Then the two companies worked together and finished the rest of the TVs in 6 days. It is known that Company B can produce 80 TVs per day. How many TVs in total did the two companies produce?

Problem 5.57 Ben and Jack can finish a task in 6 days working together. If Ben works alone for 5 days and then Jack takes over and works for 3 days, they finish $\dfrac{7}{10}$ of the work. How long would it take for each of them complete the task working alone?

Problem 5.58 A senior worker and a new worker worked together for 2 days and finished $\dfrac{3}{5}$ of their work. The senior worker then took 2 days off while the new worker continued. After that, the senior worker went back the two workers worked together to finish their work. If the senior worker works twice as fast as the new worker, how many days did it take in total to finish the work? Express your answer as a mixed number.

Problem 5.59 Calvin and Tony worked on producing a set of machines. Calvin planned to complete $\frac{7}{12}$ of the task. After he finished, he helped Tony to produce 24 pieces. The ratio of the number of pieces of Calvin to Tony is 5:3. How many pieces did Tony make?

Problem 5.60 A project is planned to be completed by 45 people, and it will take some days to do it. After 6 days of work, 9 people left the team. As a result, it takes 4 more days to complete the project than originally planned. In how many days did they originally plan to finish the project?

Answer Keys to Problems

The answer keys to the exercise problems in all the chapters are provided here. Only short answers are included. Full solutions for all problems can be found in the solutions manual, "*Mathematical Wisdom in Everyday Life Solutions Manual*".

Chapter 1. Number Sense Answer Key

Problem 1.1 $1, 3, 5, 15$

Problem 1.2 8

Problem 1.3 5

Problem 1.4 $1, 2, 3, 4, 5, 6, 8, 10.$

Problem 1.5 $\dfrac{1}{12}$

Problem 1.6 Quotient: 11, Remainder: 3

Problem 1.7 $10\frac{1}{4}$

Problem 1.8 $6\frac{1}{2}$

Problem 1.9 $\frac{33}{7}$

Problem 1.10 1

Problem 1.11 $\frac{1}{2}$

Problem 1.12 $\frac{11}{3}$

Problem 1.13 $3\frac{1}{11}$

Problem 1.14 $6\frac{1}{2}$

Problem 1.15 $\frac{3}{20}$

Problem 1.16 8

Problem 1.17 $1\frac{1}{14}$

Problem 1.18 $\frac{1}{50}$

Problem 1.19 0.35.

Problem 1.20 $\frac{23}{2}$

Problem 1.21 $1, 2, 4$

Problem 1.22 12

Problem 1.23 $4, 8, 12, 16, 20$

Problem 1.24 30

Problem 1.25 $\dfrac{4}{5}$

Problem 1.26 $\dfrac{2}{3}$

Problem 1.27 $\dfrac{12}{25}$

Problem 1.28 0.625

Problem 1.29 $\dfrac{7}{12}$

Problem 1.30 $\dfrac{3}{20}.$

Problem 1.31 $\dfrac{1}{6}$

Problem 1.32 $\dfrac{3}{4}$

Problem 1.33 $\dfrac{28}{3}$

Problem 1.34 $\dfrac{1}{8}$

Problem 1.35 $1\dfrac{1}{8}$

Problem 1.36 6

Problem 1.37 20

Problem 1.38 24

Problem 1.39 $30\,\text{cm}$

Problem 1.40 Every 24 weeks.

Problem 1.41 37

Problem 1.42 92

Problem 1.43 22

Problem 1.44 6

Problem 1.45 12 pairs of socks, $4 change

Problem 1.46 Yellow

Problem 1.47 Bob

Problem 1.48 15

Problem 1.49 225

Problem 1.50 $7\frac{1}{2}$

Problem 1.51 $5

Problem 1.52 $12\frac{5}{6}$

Problem 1.53 23

Problem 1.54 $0.\overline{2}$

Problem 1.55 $0.\overline{45}$

Problem 1.56 $0.8\overline{3}$

Problem 1.57 $\frac{2}{3}$.

Problem 1.58 $\frac{4}{15}$

Problem 1.59 $\dfrac{7}{33}$

Problem 1.60 $2\dfrac{2}{9}$

Chapter 2. Ratios, Proportion, and Percentage Answer Key

Problem 2.1 20%

Problem 2.2 153

Problem 2.3 30

Problem 2.4 30%

Problem 2.5 60%

Problem 2.6 $11.20

Problem 2.7 $73.50

Problem 2.8 20%

Problem 2.9 $86.40

Problem 2.10 $416

Problem 2.11 $123

Problem 2.12 66

Problem 2.13 25

Problem 2.14 Agent: $8100, Owner: $126900

Problem 2.15 40 cm

Problem 2.16 36

Problem 2.17 42

Problem 2.18 12

Problem 2.19 25

Problem 2.20 36

Problem 2.21 3

Problem 2.22 40%

Problem 2.23 Increase: 50%, Decrease: $\approx 33\%$

Problem 2.24 60

Problem 2.25 $2.50

Problem 2.26 $60

Problem 2.27 16

Problem 2.28 2000

Problem 2.29 4

Problem 2.30 60

Problem 2.31 27

Problem 2.32 $81

Problem 2.33 15%

Problem 2.34 $10

Problem 2.35 2 and 3

Problem 2.36 389

Problem 2.37 $134

Problem 2.38 $475.75

Problem 2.39 75%

Problem 2.40 4

Problem 2.41 $\approx 31\%$

Problem 2.42 43

Problem 2.43 35%

Problem 2.44 30%

Problem 2.45 1.4%

Problem 2.46 $39.60

Problem 2.47 44%

Problem 2.48 Medium, Large, Small

Problem 2.49 18

Problem 2.50 $1,500,000$

Problem 2.51 10

Problem 2.52 0

Problem 2.53 64

Problem 2.54 18

Problem 2.55 15%

Problem 2.56 7%

Problem 2.57 First part $120, second part $50, third part $200

Problem 2.58 5%

Problem 2.59 85%

Problem 2.60 50%

Chapter 3. Chickens and Rabbits Answer Key

Problem 3.1 26 chickens, 19 rabbits

Problem 3.2 18

Problem 3.3 40

Problem 3.4 13 packs of colored pencils, 3 packs of regular pencils

Problem 3.5 10 chickens, 6 rabbits

Problem 3.6 58 goats, 42 ducks

Problem 3.7 35 cars, 25 motorcycles

Problem 3.8 10 teachers, 89 students

Problem 3.9 4 big boats, 7 small boats

Problem 3.10 9 sets of chess, 17 sets of Chinese checkers

Problem 3.11 68 pentagons, 20 triangles

Problem 3.12 15

Problem 3.13 7

Problem 3.14 27 small, 21 big tables

Problem 3.15 55 bicycles, 35 tricycles

Problem 3.16 55 sedans, 25 minivans

Problem 3.17 26 four-person, 16 six-person teams

Problem 3.18 3 chickens, 14 goats

Problem 3.19 7 salads, 13 pizzas

Problem 3.20 5 birds, 20 cats

Problem 3.21 14

Problem 3.22 8 nickels, 8 dimes

Problem 3.23 80 birds, 20 cats

Problem 3.24 12 gallons 25% solution, 18 gallons 50% solution

Problem 3.25 $760

Problem 3.26 44 ducks, 36 sheep

Problem 3.27 $252

Problem 3.28 60 ounces

Problem 3.29 25 in City A, 15 in City B

Problem 3.30 4%: 75 pounds, 20%: 45 pounds

Problem 3.31 60%: 37.5, 100%: 62.5

Problem 3.32 20

Problem 3.33 720

Problem 3.34 80 chickens, 20 rabbits

Problem 3.35 25 dollars

Problem 3.36 30 turtles, 70 cranes

Problem 3.37 5

Problem 3.38 15 chickens, 90 rabbits

Problem 3.39 20 big containers, 30 small containers

Problem 3.40 40 three inch, 30 five inch sticks.

Problem 3.41 12 chickens, 19 rabbits

Problem 3.42 25 big mice, 75 baby mice

Problem 3.43 7 spiders, 5 fireflies, 4 cicadas

Problem 3.44 20

Problem 3.45 40 $2, 20 $5, 20 $10 bills

Problem 3.46 9 spiders, 7 dragonflies, and 4 cicadas

Problem 3.47 45

Problem 3.48 7 crabs, 8 mantises, 22 dragonflies

Problem 3.49 20 senior and 80 junior monks

Problem 3.50 20 large, 40 small

Problem 3.51 Charles 510, David 270

Problem 3.52 1250 pounds of bananas, 3750 pounds of apples

Problem 3.53 3.5 hours uphill, 2.5 hours downhill

Problem 3.54 20 red, 30 green, 30 blue

Problem 3.55 7

Problem 3.56 16 friends, 2 boats

Problem 3.57 Grape: 140, Cranberry: 60

Problem 3.58 Cashews: 2 kg, Pecans: 6 kg

Problem 3.59 Raisons: 12, Peanuts: 8

Problem 3.60 37 pounds sirloin, 63 pounds hamburger

Chapter 4. Motion—Speed, Time, and Distance Answer Key

Problem 4.1 40 miles per hour

Problem 4.2 31.5 miles

Problem 4.3 101 km/h

Problem 4.4 40 m/s

Problem 4.5 40 hours

Problem 4.6 80 km/h

Problem 4.7 $\dfrac{1300}{23}$ miles per hour

Problem 4.8 1.25 miles

Problem 4.9 48 miles per hour

Problem 4.10 4 miles per hour.

Problem 4.11 54 miles per hour

Problem 4.12 55 miles per hour.

Problem 4.13 42.5 miles/hr

Problem 4.14 40 miles per hour

Problem 4.15 12 miles

Problem 4.16 3.4 hours

Problem 4.17 8 miles

Problem 4.18 2 hours

Problem 4.19 3 km/h

Problem 4.20 20 miles

Problem 4.21 $\dfrac{105}{2}$ miles per hour

Problem 4.22 300 miles

Problem 4.23 48 miles

Problem 4.24 15

Problem 4.25 11 seconds

Problem 4.26 99 seconds

Problem 4.27 (i) 6 seconds, (ii) $\dfrac{22}{3}$ seconds

Problem 4.28 Train: 100 meters, Platform: 125 meters

Problem 4.29 4000 km

Problem 4.30 4 miles per hour

Problem 4.31 32 km/h

Problem 4.32 2 hours

Problem 4.33 3 hours

Problem 4.34 60 km/h

Problem 4.35 $\dfrac{3}{7}$ miles per minute

Problem 4.36 30 minutes

Problem 4.37 25 minutes

Problem 4.38 1488 miles

Problem 4.39 12.5 mph

Problem 4.40 60 seconds

Problem 4.41 $\dfrac{600}{19}$ cm per minute

Problem 4.42 5 hours

Problem 4.43 $\dfrac{72}{13}$ meters per second

Problem 4.44 8

Problem 4.45 2 hours 15 minutes

Problem 4.46 4 hours and 30 minutes

Problem 4.47 17.5 miles per hour

Problem 4.48 $\dfrac{550}{43}$ mph

Problem 4.49 80 meters per minute

Problem 4.50 Joe: 6 miles/hr, JoAnn: 3.6 miles/hr

Problem 4.51 Sami: 6 meters/sec, Rajan: 4 meters/sec

Problem 4.52 15 seconds

Problem 4.53 9 : 10.

Problem 4.54 9 hours

Problem 4.55 24 days

Problem 4.56 176 miles

Problem 4.57 39 mph

Problem 4.58 Amy: 20 minutes, Amanda: 30 minutes

Problem 4.59 Speed: 10 meters/sec. Length: 200 meters

Problem 4.60 294 meters

Chapter 5. Work Related Problems Answer Key

Problem 5.1 $\frac{12}{7}$ days

Problem 5.2 $\frac{6}{5}$ hours

Problem 5.3 2 hours

Problem 5.4 $\frac{40}{3}$ hours

Problem 5.5 24 minutes

Problem 5.6 $\frac{35}{6}$

Problem 5.7 $\frac{40}{13}$ hours

Problem 5.8 12 hours

Problem 5.9 5 hours

Problem 5.10 3

Problem 5.11 6.5 days

Problem 5.12 $\frac{6}{7}$ hours

Problem 5.13 2 hours

Problem 5.14 $\frac{315}{16}$ minutes

Problem 5.15 6 hours

Problem 5.16 7 hours

Problem 5.17 1500 liters

Problem 5.18 $\frac{30}{47}$ hours.

Problem 5.19 90 minutes

Problem 5.20 $\frac{4}{5}$ hours

Problem 5.21 3

Problem 5.22 $1\frac{2}{3}$ minutes

Problem 5.23 $\frac{9}{2}$ hours

Problem 5.24 2 hours: $\frac{7}{12}$, x hours: $\frac{7x}{24}$

Problem 5.25 $\frac{10}{3}$ hours

Problem 5.26 18

Problem 5.27 $\dfrac{3}{5}$

Problem 5.28 $\dfrac{41}{4}$ hours

Problem 5.29 10 days

Problem 5.30 168

Problem 5.31 $\dfrac{17}{24}$

Problem 5.32 $\dfrac{25}{2}$ hours and $\dfrac{25}{3}$ hours

Problem 5.33 20 minutes

Problem 5.34 $\dfrac{225}{2}$ minutes

Problem 5.35 $\dfrac{5}{24}$ of the pool per hour

Problem 5.36 Patty: $\dfrac{15}{4}$ hours, Tracy: $\dfrac{15}{2}$ hours

Problem 5.37 $\dfrac{1}{2}$

Problem 5.38 $\dfrac{15}{2}$ hours

Problem 5.39 168 hours

Problem 5.40 1050

Problem 5.41 A: 45 days, B: 30 days

Problem 5.42 Ethan: 8 days, Owen: 16 days

Problem 5.43 $\dfrac{1}{4}$ of a day

Problem 5.44 Iris: 96 days, Olivia: $\dfrac{32}{3}$ days

Problem 5.45 $\dfrac{7}{4}$ hours

Problem 5.46 252

Problem 5.47 50

Problem 5.48 3 hours

Problem 5.49 $3\dfrac{1}{3}$ hours

Problem 5.50 Brandon: 24 days, Richard: 40 days, Samuel: $\dfrac{120}{7}$ days.

Problem 5.51 1980

Problem 5.52 $6\dfrac{1}{4}$.

Problem 5.53 Bob

Problem 5.54 $\dfrac{60}{17}$.

Problem 5.55 76

Problem 5.56 960

Problem 5.57 Ben: 10 days, Jack: 15 days

Problem 5.58 $4\dfrac{2}{3}$.

Problem 5.59 216

Problem 5.60 22

Index

A

average speed 97

C

chicken-rabbit problem 72
common denominator 24
common factor 20
common multiple 21

D

division with remainder 10
divisor 12

F

factor 12
frame of reference 101

G

greatest common factor (GCF) 20

I

improper fraction 14

L

least common multiple (LCM) 22
lowest common denominator 26

lowest terms 23

M

meet up, farewell, and catch up 101
mixed number 14
motion in flowing water 98
multiple 10

P

percentage 43
prime 13
proper fraction 14
proportion 42

R

rate, time, distance 96
ratio 42
rational number 13
reduced fraction 24
relative speed 100

S

simplified fraction 24

W

work-related problem 122

www.ingramcontent.com/pod-product-compliance
Lightning Source LLC
Chambersburg PA
CBHW080555220326
41599CB00032B/6487